Agriculture Issues and Policies

www.novapublishers.com

Agriculture Issues and Policies

Computational Intelligence for Sustainable Development
Brojo Kishore Mishra, PhD (Editor)
2022. ISBN: 979-8-88697-198-9 (Hardcover)
2022. ISBN: 979-8-88697-346-4 (eBook)

Pest Management: Methods, Applications and Challenges
Tarique Hassan Askary, PhD (Editor)
2022. ISBN: 979-8-88697-268-9 (Hardcover)
2022. ISBN: 979-8-88697-393-8 (eBook)

Strategies to Achieve Sustainable Development Goals (SDGs): A Road Map for Global Development
Rajani Srivastava, M.Sc., PhD (Editor)
2022. ISBN: 978-1-68507-836-2 (Hardcover)
2022. ISBN: 979-8-88697-027-2 (eBook)

Pistachios: Cultivation, Production and Consumption
Shaziya Haseeb Siddiqui, PhD
Shoaib Alam Siddiqui, PhD (Editors)
2022. ISBN: 978-1-68507-949-9 (Hardcover)
2022. ISBN: 979-8-88697-063-0 (eBook)

Ecosystem Services: Types, Management and Benefits
Hanuman Singh Jatav, PhD
Vishnu D. Rajput, PhD (Editors)
2022. ISBN: 978-1-68507-614-6 (Hardcover)
2022. ISBN: 978-1-68507-747-1 (eBook)

More information about this series can be found at
https://novapublishers.com/product-category/series/agriculture-issues-and-policies/

Matthieu Issa
Editor

Jute

Cultivation, Properties and Uses

Copyright © 2023 by Nova Science Publishers, Inc.

All rights reserved. No part of this book may be reproduced, stored in a retrieval system or transmitted in any form or by any means: electronic, electrostatic, magnetic, tape, mechanical photocopying, recording or otherwise without the written permission of the Publisher.

We have partnered with Copyright Clearance Center to make it easy for you to obtain permissions to reuse content from this publication. Simply navigate to this publication's page on Nova's website and locate the "Get Permission" button below the title description. This button is linked directly to the title's permission page on copyright.com. Alternatively, you can visit copyright.com and search by title, ISBN, or ISSN.

For further questions about using the service on copyright.com, please contact:
Copyright Clearance Center
Phone: +1-(978) 750-8400   Fax: +1-(978) 750-4470   E-mail: info@copyright.com.

NOTICE TO THE READER

The Publisher has taken reasonable care in the preparation of this book, but makes no expressed or implied warranty of any kind and assumes no responsibility for any errors or omissions. No liability is assumed for incidental or consequential damages in connection with or arising out of information contained in this book. The Publisher shall not be liable for any special, consequential, or exemplary damages resulting, in whole or in part, from the readers' use of, or reliance upon, this material. Any parts of this book based on government reports are so indicated and copyright is claimed for those parts to the extent applicable to compilations of such works.

Independent verification should be sought for any data, advice or recommendations contained in this book. In addition, no responsibility is assumed by the Publisher for any injury and/or damage to persons or property arising from any methods, products, instructions, ideas or otherwise contained in this publication.

This publication is designed to provide accurate and authoritative information with regard to the subject matter covered herein. It is sold with the clear understanding that the Publisher is not engaged in rendering legal or any other professional services. If legal or any other expert assistance is required, the services of a competent person should be sought. FROM A DECLARATION OF PARTICIPANTS JOINTLY ADOPTED BY A COMMITTEE OF THE AMERICAN BAR ASSOCIATION AND A COMMITTEE OF PUBLISHERS.

Additional color graphics may be available in the e-book version of this book.

Library of Congress Cataloging-in-Publication Data

ISBN: 979-8-88697-490-4

Published by Nova Science Publishers, Inc. † New York

Contents

Preface		 vii
Chapter 1	**Manufacturing Processes of Jute Fiber-Based Composites**..1 Edgar A. Franco-Urquiza	
Chapter 2	**Chemical Modifications of Jute Fiber Properties for Lifecycle Enhancement by Utilizing in Wastewater Treatment** ....51 Aleksandra Ivanovska and Mirjana Kostic	
Chapter 3	**Progress, Challenges, and Prospects of Jute Fiber as Green Adsorbents: A Scope Beyond Traditional Applications** ....95 Aparna Roy	
Chapter 4	**Jute Fiber: Extraction, Properties and Applications** ....111 Md. Vaseem Chavhan, B. Venkatesh and Beera Murali	
Chapter 5	**Jute: Potential Applications in Projects of Environmental Recovery and Environmental Conservation** ....127 L. H. Tsuchiya and A. M. Da Silva	
Index		141

Preface

This volume contains five chapters discussing the cultivation, properties, and uses of jute fiber. Chapter One is an overview of the manufacturing processes of jute fiber-based composites. Chapter Two discusses methods of chemical modification of the properties of jute fiber for lifecycle enhancement in wastewater treatment. Chapter Three examines in a scope beyond traditional applications the progress, challenges, and prospects of jute fiber as a green adsorbent. Chapter Four provides a review of the processing of jute fiber from extraction to products. Chapter Five discusses potential applications of jute in projects of environmental recovery and environmental conservation.

Chapter 1 - This book chapter reviews the characteristics of jute fibers that allow their interaction with polymer resins for the formation of composite materials. Jute fibers used to manufacture products and fabrics are made up of several individual filaments forming fiber rovings to offer more excellent mechanical performance. Jute has several weaknesses: it is flammable, degrades thermally, and is highly hygroscopic. Therefore, it is necessary to carry out fiber treatments to modify it to be used in developing sustainable products. The main disadvantages provided by the intrinsic nature of natural fibers in combination with polymeric matrices are poor fiber-matrix compatibility and relatively high moisture absorption. Polymer matrices can be thermoplastic or thermosetting. Thermoplastic polymers can be melted down and reprocessed multiple times, making them recyclable. Thermosetting resins are the most used in high-performance composite materials. These resins give rise to a rigid, insoluble, and infusible product through a series of chemical reactions (curing or cross-linking). Different techniques are used to manufacture polymer composites depending on the configuration and size of the fibers and the type of polymer matrix.

Chapter 2 - The increased demand for cheap, biodegradable, renewable, and recyclable fibers with exceptional properties positioned jute (*Corchorus capsularis* L. and *Corchorus olitorius* L.) in the second place (after cotton) in the natural fiber world market. Multicellular jute fibers are comprised of three main components: cellulose, lignin, and hemicelluloses having a wide variety of functional groups capable of binding different water pollutants. However, such groups are not easily accessible due to the presence of a hydrophobic surface layer (consisting of pectins, waxes, and fats) that could be removed by applying simple alkali and oxidative modifications. Moreover, fibers' activation using different chemical agents or grafting of functional groups on their surfaces results in enhanced fiber sorption properties, and hence adsorption potential for various water pollutants. This chapter provides an overview of the possibility of the application of raw and chemically modified jute fibers as an eco-friendly adsorbent for heavy metals and dyes as the most frequent water pollutants. Special attention has been paid to the binding mechanism of the pollutants and differently functionalized jute adsorbents. The last section of this chapter represents one step toward both the circular economy approach and sustainable development, in terms of reusing and revalorization of solid waste with adsorbed pollutants. Permanent collection and reuse of pollutant saturated jute adsorbents have promising multi-positive effects on the economy as well environment, including reducing its quantity, saving energy, and its utilization as raw material for producing new hybrid materials which is in line with the Circular Economy Package (2020).

Chapter 3 - Currently, water crisis and pollution and its management and possible solutions are recognized as a distinctive challenge faced by humankind. The major contributors to water pollution are discharging of industrial effluents in water bodies, contaminated with different heavy metal ions, dyes, hydrocarbons and other harmful chemicals. Though different techniques including photocatalytic decolorization and oxidation, biological degradation, coagulation and precipitation, ion exchange, membrane filtration, etc. are the conventionally used pollutant removal procedures, adsorption is one of the most popular and fundamental processes for wastewater treatment and water reclamation. Nowadays, among a large variety of adsorbents, activated carbon is one of the commericialized adsorbent for the wastewater treatment. But the adsorbent grade activated carbons are rather expensive and its usages are also associated with the difficulties of subsequent treatment, regeneration and disposal of the spent carbon. These constraints rendered the researchers to find a simple, economic and efficient adsorbent for pollutant removal from wastewater. From the view of environmental issues, in the search of alternative and inexpensive adsorbent, the efforts were mainly focused on the biological materials. Several experiments were conducted with a wide variety of biomaterials to investigate their feasibility as adsorbents. However, most of these are unavailable plentifully in the global market, which eventually makes them incapable to meet the huge commercial demand, and maximum are also not highly efficient enough to be applied to real industrial wastewater. Thus, the exploration of a novel, environment friendlier, easily available, cheap and effective adsorbent for wastewater treatment, is still necessary. Regarding the current scenario, this review explores the feasibility of the novel application of abundantly available lignocellulosic jute fiber as a potential bioadsorbent for wastewater. Jute, the second most abundantly available natural bast fiber, is primarily comprised of cellulose (64.4%), hemicellulose (12%) and lignin (11.8%). Today, jute industry, a vital sector of South-East Asia, is critically challenged with several major threats to survival. Thus, widening the scope of jute fiber by utilizing it as bioadsorbent may offer a sustainable technology for wastewater treatment as well as promote the jute industries, which will be in turn beneficial for the jute farmers.

Chapter 4 - The jute fibre (*Corchorus olitorius*) belongs to the category of natural cellulosic bast fibre. The fibre cells are separated from the bast by the retting process to extract the individual fibres. The different media, moisture, water, chemical, and enzymes are used for the retting process based on the time of the process and the quality of fibre to be obtained. The extracted fibre is further processed to obtain the finished fibre of the required fineness and luster. Jute fibers are characterized by their unique multicellular structure, having cellulosic microfibrils covered with lignin. The jute fibre is known for its good initial modulus, dimensional stability, toughness, and antimicrobial properties among cellulosic natural fibres. Other than these properties, the low cost of fibre, and the world's focus towards sustainability, the jute fibre is gaining its scope in various applications like fibre reinforced composites, geotechnical, automotive, and construction applications. In the present chapter, the jute fibre cultivation, and extraction by retting have been discussed. Further, the conversion of extracted fibre to yarn and fabric is explained in detail.

Chapter 5 - In this work, the authors research and present some biological and ecological properties of jute. The authors also present a summarized set of case studies and they performed a PESTLE analysis to facilitate the presentation of some results. The authors have found that there is a vast potential for the use of jute-based products, both as fibers and as live plants, featuring an open range of business opportunities and new research. Regarding the potential for new scientific works, the authors address about the necessity to improve the production method, in order to make the production system truly sustainable. Also, there is a demand to use fibers in their most natural state as possible, because although jute fiber is biodegradable (perhaps its best ecological property), the fact of applying synthetic products to improve the quality of the product can make it as harmful to ecosystems as synthetic fibers.

Chapter 1

Manufacturing Processes of Jute Fiber-Based Composites

Edgar A. Franco-Urquiza[*], PhD
Advanced Manufacturing Department,
Center for Engineering and Industrial Development (CIDESI), Querétaro, México

Abstract

This book chapter reviews the characteristics of jute fibers that allow their interaction with polymer resins for the formation of composite materials. Jute fibers used to manufacture products and fabrics are made up of several individual filaments forming fiber rovings to offer more excellent mechanical performance. Jute has several weaknesses: it is flammable, degrades thermally, and is highly hygroscopic. Therefore, it is necessary to carry out fiber treatments to modify it to be used in developing sustainable products. The main disadvantages provided by the intrinsic nature of natural fibers in combination with polymeric matrices are poor fiber-matrix compatibility and relatively high moisture absorption. Polymer matrices can be thermoplastic or thermosetting. Thermoplastic polymers can be melted down and reprocessed multiple times, making them recyclable. Thermosetting resins are the most used in high-performance composite materials. These resins give rise to a rigid, insoluble, and infusible product through a series of chemical reactions (curing or cross-linking). Different techniques are used to manufacture polymer composites depending on the configuration and size of the fibers and the type of polymer matrix.

Keywords: jute fiber, manufacturing processes, natural fibers, green composites

---

[*] Corresponding Author's Email: edgar.franco@cidesi.edu.mx.

In: Jute: Cultivation, Properties and Uses
Editor: Matthieu Issa
ISBN: 979-8-88697-490-4
© 2023 Nova Science Publishers, Inc.

Introduction

Jute is one of the most accessible natural fibers and is considered the second natural fiber after cotton. Jute is long, soft, and shiny, with a length of 1 to 4 m and a diameter of 17 to 20 microns. Jute fibers are composed primarily of plant materials such as cellulose and lignin, and they are biodegradable and recyclable. Jute fibers have low cost, high abundance, and reasonable mechanical properties. Some technical advantages of jute fiber are low-cost manufacturing and lightweight with optimal mechanical properties. Various factors related to fiber gauge and configuration (non-woven and woven), fiber layout, nature of polymer matrices, and composite manufacturing techniques influence its performance. Despite the general advantages, the fiber has several drawbacks, such as relative roughness, brittleness, wide variation in fiber length and fineness, branched nature, and a tendency to yellow under sunlight. The intrinsic nature of natural fibers provides the main disadvantages in combination with polymer matrices. Molecular incompatibility between natural fiber reinforcements and polymeric matrices affects the interaction of the constituents, resulting in a poor fiber-matrix interface. Combining the hydrophilic jute fiber with the hydrophobic polymer matrix affects mechanical efficiency due to the weak interface that reduces stress transfer from the matrix to the fibers. Thermoplastic and thermoset resins are used as polymer matrices in developing jute fiber composites. Both kinds of resins can be synthetic or biobased resins that contribute to the performance of composites, including the environment, through a circular economy. Several manufacturing processes, including combined technologies, are used to fabricate jute fiber composites. The demand for fiber-reinforced composite materials is increasing in structural applications due to their crucial characteristics, such as stiffness, strength and durability, and low-cost processing benefits.

Jute Fiber

Literature Review

Jute is a plant belonging to the Tiliaceae family of the botanical genus *Corchorus*, with around 40 species distributed throughout the tropics. According to Atkinson (Anon n.d.), the most widely used for trade are *Corchorus capsularis*, also known as 'white' jute, and *Corchorus olitorius*, known as 'Tossa' jute. Jute fiber is one of the most used worldwide. India, Bangladesh, China, Nepal, Myanmar, and Thailand are the leading natural fiber producers (Kundu 1959). According to the Food and Agriculture Organization (FAO), India accounts for more than 50% of the world's jute production. In America, jute cultivation is favored by the climatic region of the Amazon zone. For the cultivation of jute, a series of conditions must be met, among which a warm and humid climate stands out since it needs abundant water during its growth phase. The jute plant can reach a height of about 3-4 meters and a thickness of about two centimeters in diameter(Adak and Mukhopadhyay 2016; Chauhan, Kärki, and Varis 2019; Franco-Urquiza 2022; Jagadish and Sumit 2021; Mathur 2021; Rangappa et al. n.d.).

From the entire jute plant, the part from which the fiber is extracted is from the phloem vessels located below the main stem. Jute fiber consists of several polygonal unit cells, each of which has a central lumen, primary and secondary cell walls, as well as intermediate lamellae (Franco-Urquiza 2022; Mathur 2021; Rangappa et al. n.d.). The jute fibers or filaments are gradually separated during the carding process to obtain a spinning fiber or a single jute filament of variable length. The maximum length of a jute filament is about 30 cm. Various studies describe a variable number of cells (Anon n.d.; Kundu 1959; Mathur 2021), which affect the strength of the fiber. The fineness of jute fiber is highly variable and is usually 2-2.5 tex. Thus, a jute thread that has 70-80 fibers in cross section is approximately 140-210 tex (4-6 lbs. /spy).

Jute fiber is used for decoration products, upholstery, packaging, and other textile applications, including technical textiles, due to its relevant physical and mechanical properties. Jute fiber has been attracting the industry's attention in recent years for its use in non-structural applications because it is an agro-renewable fiber, biodegradable, and relatively available at a low price. Jute has relevant technical properties such as high tensile strength, dimensional stability, and good acoustic and thermal insulation (Rangappa et al. n.d.; Wang et al. n.d.)

Despite the general advantages, the fiber has several drawbacks, such as relative roughness, brittleness, wide variation in fiber length and fineness, branched nature, and a tendency to yellow under sunlight. According to several authors (Chandekar, Chaudhari, and Waigaonkar 2020; Chauhan et al. 2019; Das and Chaudhary 2020; Elanchezhian et al. 2018; ElayaPerumal and Venkateshwaran 2008; Gon et al. 2012; Rana and Jayachandran 2000; M. Torres-Arellano, Renteria-Rodríguez, and Franco-Urquiza 2020; Mauricio Torres-Arellano, Renteria-Rodrígucz, and Franco Urquiza 2020), jute fiber performance varies due to natural variability in internal and surface microstructural characteristics, which some factors can influence, including growing conditions (i.e., temperature, humidity, soil conditions), quenching (aqueous, dew, or enzymatic) and fiber extraction processes, fiber length and diameter, chemical constituents and their proportional amounts. Microstructural characteristics contribute significantly to the fiber's physical, mechanical, and thermal performance. For this reason, it is necessary to make chemical modifications to the fibers to extend their use in the manufacture of technical textiles and of high added value in developing polymeric composites reinforced with jute fiber (Balla et al. 2019; Chandekar et al. 2020; chubuike et al. 2017; Khan and Alam 2016; Sathishkumar et al. 2017; Wang et al. n.d.).

Jute Fiber Characterization
In the previous section, the general morphology of jute fiber was described, highlighting the non-homogeneity of the fiber, which provides a wide variation of its properties. The morphological studies commonly carried out on jute fiber and the evaluation of its mechanical performance in filament, thread, and fabric (non-woven and woven) will be briefly presented in the following sections.

Jute fiber morphology is generally characterized using scanning electron microscopy (SEM). One of the first works presenting the characterization of jute fiber by SEM was published approximately four decades ago. Guha Roy et al., (Guha Roy, Mukhopadhyay, and Mukherjee 1984) studied the surface characteristics of jute fiber after various degrees of delignification and bleaching treatments. They found that the surface characteristics change progressively with the gradual removal of lignin from the jute fiber at 93% delignification. Degradation or damage to fibers or cells occurs when the chlorine used is near neutral or acidic. The photographs of the different fibers obtained from SEM revealed the changes in the surface characteristics of these fibers. These changes could be explained by the gradual increase in delignification of jute fibers.

More recently, the fiber microstructure greatly influences fiber properties. One way to investigate the microstructure of jute fiber is by using low-voltage scanning electron microscopy (LV-SEM). Sameer F. Hamad et al. (Guha Roy et al. 1984) characterized the microstructure of different natural plant fibers (flax, jute, ramie, and sisal fibers) using the LV-SEM technique. LV-SEM observations indicated that jute and sisal fibers exhibit less variation in fiber cross-sectional area, internal lumen shape and size, and cell wall thickness than sisal fibers, flax, and ramie. The authors found that the morphology of natural fibers is also reflected in the mechanical properties of the fibers.

Observations in LV-SEM revealed that primary cell wall collapse generally leads to a non-linear stress-strain curve for individual fibers. The actual cross-sectional area of individual fibers was measured by LV-SEM image analysis using image software J. Saulo Rocha Ferreira et al. (Ferreira et al. 2016) evaluated the bonding behavior of jute fibers in a cement-based matrix. Figure 1 represents some examples of the microstructural analysis of the jute fibers used in their study. The authors found that an irregular shape characterizes the cross-sectional geometry of jute fiber. Also, the size of the cross-section varies along the fiber.

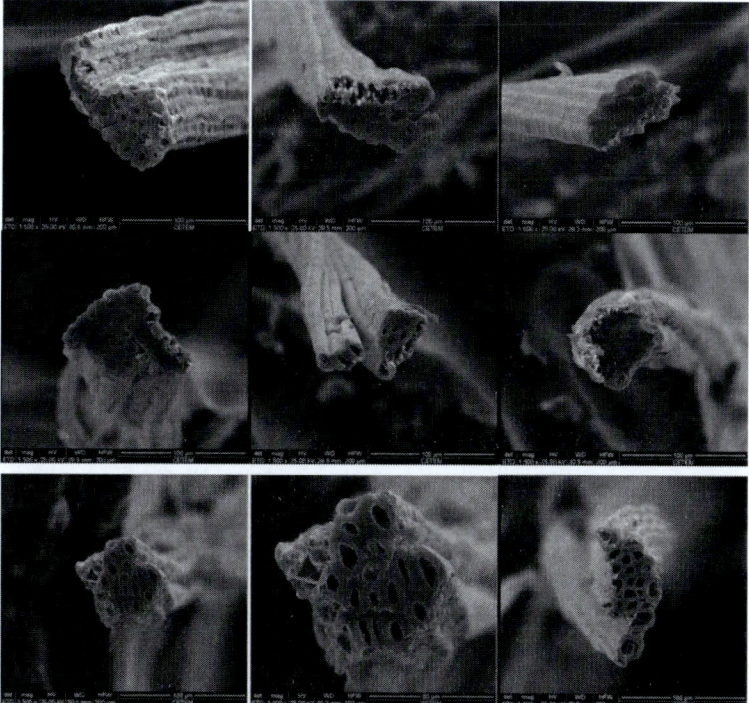

Figure 1. SEM micrographs corresponding to the cross-section morphology of jute fibers (Ferreira et al. 2016).

Mechanical Performance of Jute Fibers

Jute fibers used to manufacture products and fabrics are made up of several individual filaments forming fiber rovings to offer more excellent mechanical performance. However, it is of genuine interest to know the mechanical properties of the individual filaments of jute fiber.

Shahinur, S. et al. (Shahinur et al. 2015) cut a long jute fiber into three different portions for characterization. The authors highlighted that the middle portion had better mechanical, thermal, chemical, and crystalline properties than the other two jute fiber portions. The diameter of the jute fiber turned out to be gradually thinner at the ends due to the variation of its maturity. The top grain was immature, the middle grain was correctly ripened, and the cut part was overripe. There were fewer pores and voids in the fiber surface of the upper portion compared to the surfaces of the middle and lower portions (also known as the shear zone). The cut portion's surface was rougher than the upper and middle portions, as presented in Figure 2. The surface morphology of the different portions of the jute fiber was found to be different due to their variations in diameter and maturity. Fewer micropores in the surface morphology generally indicate immature fiber.

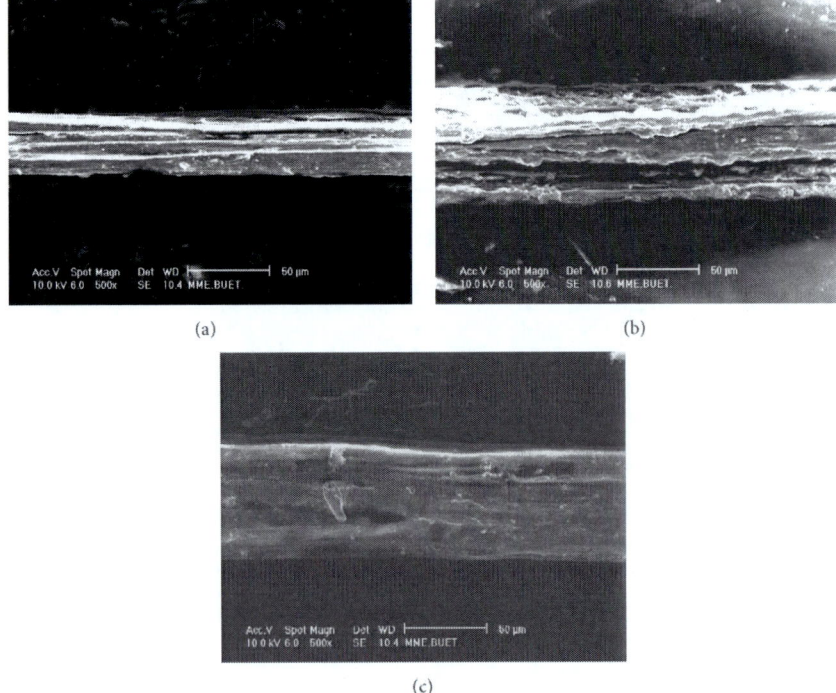

Figure 2. SEM micrographs corresponding to the surface morphology of: (a) top, (b) middle, and (c) cutting portions of raw jute fiber (Shahinur et al. 2015).

Therefore, it is considered that the middle portion of the jute fiber has the best mechanical, thermal, chemical, and crystalline properties compared to the extreme parts of the fiber.

Biswas et al. (Biswas et al. 2011) evaluated the effect of span length on the tensile properties of jute fiber from Bangladesh. They stacked jute fiber with different section lengths strands between two paper frames to form a good grip on the grips of a universal testing machine and provide linear steering during testing. Young's modulus and strain at break were corrected using analytical equations recently developed by the authors to correlate both mechanical parameters. The correction method resulted in a high Young's modulus for larger spans, while the strain at break was lower than smaller ones. The longer span length helps to minimize machine travel compared to smaller ones. Through SEM observations, a smooth surface and compact structure were observed, which determined that Young's modulus increases with the increasing length of the section. In contrast, the tensile strength and breaking stress decreased with an increase in the length of the jute fiber stretch.

The literature consulted (Alkbir et al. 2016; Drzal n.d.; Franco-Urquiza 2022; Franco-Urquiza et al. 2020; Franco-Urquiza and Rentería-Rodríguez 2021; Edgar Adrián Franco-Urquiza, Saleme-Osornio, and Ramírez-Aguilar 2021; Guha Roy et al. 1984; Khan and Alam 2016; Kundu 1959; Rajesh and Prasad 2014; Ramamoorthy, Skrifvars, and Persson 2015; Salman 2020; Samanta, Mukhopadhyay, and Ghosh 2020; Sathishkumar et al. 2017; Shahinur et al. 2015, 2022; M. Torres-Arellano et al. 2020; Torres et al. 2020, 2020; Wang et al. n.d.) points out that jute, like other natural fibers, has high variation, indicating uncertainty in obtaining fibers with consistent mechanical properties. The way jute is mechanically tested could influence the distribution of its mechanical properties. These variations can be evaluated through statistical distribution. The Weibull distribution (Defoirdt et al. 2010) estimates a probability based on measured data and is often used to describe the strength of fiber-reinforced polymer (FRP) composites. The two-parameter Weibull distribution is typically used (Zhang et al. 2002), although the three-parameter Weibull distribution is more robust and may better characterize the data. Saaidia et al. (Saaidia et al. 2015) evaluated the mechanical properties of jute threads using a reference length of jute fiber of 50 mm. Due to the natural variability of natural fibers, 320 samples were tested at 2 mm/min, room temperature, and hygroscopicity of approximately 55%. The authors analyzed the results using a two- and three-parameter Weibull distribution by Minitab version 16 software. The Weibull distribution showed high dispersion in the mechanical properties, which increase as the number of samples increases, where they become constant from the 200 samples tested. The two-parameter Weibull gave results close to those obtained experimentally, while those obtained by the three-parameter Weibull are far from the experimental ones.

In order to produce the best quality jute products, the main requirement is to ensure the quality of the jute yarn, which can be defined through the mechanical parameters of strength and Young's modulus. However, producing good quality yarn considering these parameters is difficult to achieve because these parameters follow a non-linear relationship (Paul et al. 2022).

Jute has several weaknesses: it is flammable, degrades thermally, and is highly hygroscopic. Therefore, it is necessary to carry out fiber treatments to modify it to be used in developing sustainable products. Various modifications have been suggested in the literature to improve the functional performance of jute fiber used as reinforcement in composite materials (Chandekar et al. 2020; chubuike et al. 2017; Guha Roy et al. 1984; Hossen et al. 2020; Khan and Alam 2016; Sathishkumar et al. 2017; Saw and Datta 2009; Shahinur et al. 2020; Viju and Thilagavathi 2022; Wang et al. n.d.).

Functional Treatment of Jute Fiber

The main disadvantages provided by the intrinsic nature of natural fibers in combination with polymeric matrices are poor fiber-matrix compatibility and relatively high moisture absorption. Molecular incompatibility between natural fiber reinforcements and polymeric matrices affects the interaction of the constituents, resulting in a poor fiber-matrix interface. Combining the hydrophilic jute fiber with the hydrophobic polymer matrix affects mechanical efficiency due to the weak interface that reduces stress transfer from the matrix to the fibers. This prevents the potential of jute fiber as a reinforcing material in a composite material. In order to improve the compatibility between the fiber and the polymer matrix, the jute fiber must be physically or chemically modified. As a complement to SEM observations, jute fiber treated is generally evaluated using X-ray diffraction (XRD) and Fourier Transform Infrared (FTIR) spectroscopy. As stated, extensive literature detailing natural fibers' chemical surface modification (Chandekar et al. 2020; chubuike et al. 2017; Guha Roy et al. 1984; Hossen et al. 2020; Khan and Alam 2016; Sathishkumar et al. 2017; Saw and Datta 2009; Shahinur et al. 2020; Viju and Thilagavathi 2022; Wang et al. n.d.). Among the widely reported chemical modification techniques, alkali treatments (Adak and Mukhopadhyay 2016; Das and Chaudhary 2020; Rajesh and Prasad 2014; Ray et al. 2004; Roy, Bag, et al. 1991; Roy, Sen, et al. 1991; Samanta et al. 2020; Sathishkumar et al. 2017; Saw and Datta 2009; Shivamurthy et al. 2020), bleaching (Guha Roy et al. 1984; Khan and Alam 2016; Roy, Bag, et al. 1991; Roy, Sen, et al. 1991; Samanta et al. 2020), acetylation (Chandekar et al. 2020; Das and Chaudhary 2020; Guha Roy et al. 1984; Rangappa et al. n.d.) and graft copolymerization (Khan and Alam 2016; Miah, Khan, and Khan 2011; Viju and Thilagavathi 2022) stand out. These treatments reduce the intrinsic hydrophilicity of natural fibers (Viju and Thilagavathi 2022). Among all reported methods, sodium hydroxide treatment is highly effective in removing lignin, pectin, and hemicellulose from natural fibers (Franco-Urquiza 2022; Rangappa et al. n.d.). Other authors reported that alkaline treatment could improve mechanical interlocking and surface roughness and increase the fiber's cellulose content (Drzal n.d.; Miyagawa et al. 2006). The alkaline treatment also increases tensile strength and reduces the water absorption of fibrous materials (E. A. Franco-Urquiza, Vázquez, and Escalante-Velázquez 2021; Sathishkumar et al. 2017; Shivamurthy et al. 2020). One of the modification techniques used to remove lignin is the bleaching of cellulosic fibers since the presence of surface impurities -such as pectin and the waxy substance of the fiber (minor constituents)- sometimes hinders the process of modification. Therefore, surface pretreatment is highly desired before the chemical modification of natural fibers (Guha Roy et al. 1984; Viju and Thilagavathi 2022; Wang et al. n.d.). Sodium Chlorite bleaching is a very effective pre-treatment method to remove wax, pectin, and lignin (Roy, Bag, et al. 1991; Roy, Sen, et al. 1991).

Roy et al. (Roy, Bag, et al. 1991), bleached jute fiber with sodium chlorite and alkaline hydrogen peroxide solutions. Infrared spectra of bleached samples and jute were analyzed and compared. Bleached samples were characterized by higher absorbance intensity ratios of bands attributed to hemicellulose. Bands attributed to lignin are absent or very weak in chlorite-bleached jute compared to peroxide-bleached jute, although some residual lignin was detected in the substrate. The loss of lignin in the chlorite-bleached sample involves the oxidative degradation of the aromatic rings, while in the case of the peroxide-bleached sample, such losses are negligible due to oxidations limited to specific chromophore groups. The vibration of water molecules adsorbed on the non-crystalline regions of the cellulose appears as a sharp peak in the chlorite-bleached jute and as a shoulder in the peroxide-bleached jute bar. The authors attributed this result to the difference between the two bleaching processes.

Nayak et al. (Nayak and Mohanty 2019) reported that sodium chlorite treatment reduced the water absorption capacity of cellulosic fibers. In a recent study (Guo, Sun, and Satyavolu 2019), the peroxide treatment improved kenaf fibers' tensile properties and cellulose content. Several researchers studied the potential use of cellulosic bast fibers such as jute, flax, hemp, ramie, and kenaf in composite applications (Ashok, Srinivasa, and Basavaraju 2018; Drzal n.d.; ElayaPerumal and Venkateshwaran 2008; Franco-Urquiza 2022; E.A. Franco-Urquiza, Saleme-Osornio, and Ramírez-Aguilar 2021; Parbin et al. 2019; Puglia, Biagiotti, and Kenny 2005; Taylor, Karus, and Kaup 2002; Thyavihalli Girijappa et al. 2019; Mauricio Torres-Arellano et al. 2020; Torres et al. 2020; Westman et al. 2010).

In the case of jute fiber, chemical modification is desired due to the existence of hydroxyl groups. The objective of the modification is to use the hydrogen bonds within the cellulose molecules to activate the OH groups and obtain efficient bonds for the molecular interaction of the constituents.

Jute Fiber-Based Composites

Polymer Matrix

Polymer matrix composites are used in various products due to their lightness and specific properties. The transformation processes of polymer composites save cost and resources.

The reinforcements can be particles or fibers, where the distribution and orientation are of great interest in the development of composites. In the case of fibers, these can be large and arranged in a unidirectional way, or they can have different weave configurations to induce the properties of the composite. Fiber-reinforced composite materials are the most important from a technological point of view. Its objective is to obtain highly resistant materials to fatigue and rigidity at low and high temperatures. At the same time, a low density is sought, which is why it is intended to achieve a better strength-to-weight ratio.

Polymer matrices can be thermoplastic or thermosetting. Thermoplastic polymers can be melted down and reprocessed multiple times, making them recyclable. Thermosetting resins are the most used in high-performance composite materials. These resins give rise to a rigid, insoluble, and infusible product through a series of chemical reactions (curing or cross-linking). Different techniques are used to manufacture polymer composites depending on the configuration and size of the fibers and the type of polymer matrix.

Thermoplastic Matrix

Thermoplastic polymers can deform or become flexible at temperatures above the glass transition temperature, so they have high recyclability. Based on their monomer, plastics can be classified into at least five groups or families: polyolefins, polystyrenes, polyenes, polyvinyls, and polyacrylics.

Low and high-density polyethylene (LDPE/HDPE), polyvinyl chloride (PVC), polystyrene (PS), polypropylene (PP), Polyamide (PA), and polyacrylic (PMMA) are commonly used thermoplastics. Some biodegradable matrices, such as polylactic acid (PLA), have broad environmental advantages and have attracted attention for research and technological development in recent years. Thermoplastics are commonly processed by compression molding, extrusion, injection molding, and vacuum forming. For the manufacture of thermoplastic composites, jute fibers in the form of particles, short fibers, fabrics, and random MAT can be used depending on the applications and performance requirements.

Polypropylene is widely used due to its low cost, low thermal expansion, and recyclability (Chandekar et al. 2020). Physical and chemical modifications can improve the adhesion between hydrophilic jute fibers. The modification involves the use of plasma, steam, and ionizing radiation. In contrast, chemical treatment includes alkali, acetylation, maleate coupling agents such as PP grafted with maleic anhydride, and silane coupling agents (Chandekar et al. 2020).

Thermoset Matrix

Thermosetting polymers, such as epoxy, phenols, polyester, and vinyl ester, come in liquid form at room temperature and are mainly obtained from non-renewable resources. These resins are relatively easy to obtain and not too expensive. During the curing process, cross-linked structures are formed, which provide outstanding mechanical properties, in particular strength and stiffness, while ductility is low. Once the curing process is complete, these resins cannot be reheated, recycled, or reused, nor can they biodegrade.

Currently, great efforts are being made to develop bio-based thermosetting polymers from vegetable oils, natural phenolic complexes, and other natural sources that could have better thermal stability than thermoplastic matrices, to obtain a sustainable thermosetting, which can contribute to reducing the carbon footprint (Franco-Urquiza et al. 2020; Pawar et al. 2016; Thiagamani et al. 2019; Mauricio Torres-Arellano et al. 2020; Yorseng et al. 2020).

Conventional Manufacturing Processes

Generally, unidirectional, woven, and MATs are impregnated with thermosetting resins to form Jute Fiber Reinforced Polymer (JFRP) composites. The resin mainly used for developing these composites are epoxy and polyester resins. These composites can be processed by manual rolling, spraying, vacuum resin infusion, resin transfer molding, and compression. Other high technologies like autoclaves are used for pre-impregnated fibers.

The processing techniques are selected among multiple factors such as the type of polymer, the configuration of processing continuous or non-continuous, the mold capacity (open or closed), the presentation of the fiber (power, short, long, woven, etc.), the hybridization of fibers, the quality of the final product, the processing time, the purpose of the composite and the budget for its manufacture.

FRP Composites

FRP composites consist of a thermosetting matrix reinforced with fibers, mainly woven, unidirectional, or MATs, which results in a material with high mechanical performance and low density. These qualities make it suitable for a wide range of structural applications. There is a vast field of research regarding FRP composites. The arrangement of the fibers, the configuration of the fabrics, the combination of materials, and the different manufacturing processes are just some of the variables that researchers will have to deal with in the coming decades.

Hand Lay-Up

In the hand lay-up or manual rolling process, the mold is coated with a non-stick wax for easy removal. Subsequently, the first layer of gel-coat is applied, a component based on pigmented resin and styrene, mixed with a catalyst that favors the polymerization reaction. This first layer will finally be the visible part of the manufactured piece. When the gel-coat is completely dry, the lamination stage begins, depositing successive layers of fiber fabrics and thermosetting resin. The resin is spread using a spatula on the fabric to homogeneous impregnation. The spatula can also be used to remove the excess poured resin. The final stage consists of demolding the composite and finishing the piece, using techniques such as polishing, machining, and assembly. This manufacturing process is generally performed for the manufacture of large parts.

Spray lay-up molding is not that different from manual laying. Spray lay-up is an open mold composite manufacturing process where resin and reinforcements are sprayed into a reusable mold. Resin and fiber can be applied separately or simultaneously by a combined flow from a resin metering gun and fiber cutting device. Subsequently, the fiber and resin laminate formed on the mold can be manually compacted with rollers. The part is cured at room temperature or by heating, cools, and is removed from the mold (Nurul Hidayah et al. 2019).

Kumar et al. (Kumar and Srivastava 2017) prepared epoxy resin specimens and jute fiber/epoxy composite specimens using the manual lamination technique. The authors highlighted that the composite exhibited better tensile and compressive strength than the epoxy resin. Fiber bundle strength decreases with an increasing number of fibers in a bundle (Thygesen et al. 2011). The impact strength has no noticeable change after fiber addition. The authors highlight that these composites can be very useful for construction purposes, partitions, wall frames, floors, windows and doors, roof tiles, and mobile or prefabricated buildings that can be used in times of natural calamities such as floods, cyclones, and earthquakes (Kumar and Srivastava 2017).

The hand lay-up process allows the reinforcement of JFRP compounds by adding nanoparticles. Thus, the performance and functionality of natural fiber-reinforced polymer composites can be improved by incorporating particles that can significantly increase mechanical properties. Vivek et al. (Singh, Kumar, and Srivastava 2022) developed a study to hybridize jute fiber, and epoxy resin (JFRP) composites with cement particles to improve their mechanical properties. Jute fabric reinforced polymer filled with laminated cement particles (CmJFRP) and JFRP composites were manufactured by hand lay-up. Tensile tests revealed that CmJFRP composites exhibit lower tensile strength than JFRP composites, mainly attributed to the agglomeration of cement particles at the fiber/matrix interface. Microhardness, impact strength, and critical buckling load were higher for CmJFRP composites than JFRP composites. SEM observations revealed that the crack front changed plane and direction when it encountered the cement particles.

Vacuum Assisted Resin Transfer Molding

Vacuum Assisted Resin Transfer Molding (VARTM), also knowns as infusion process, emerged as an evolution of the hand lay-up method, seeking to solve the problems of low reinforcement content in its composition. The process is carried out using a plastic bag placed over the laminate. The air inside the bag is extracted employing a vacuum system, thus achieving pressures of almost one atmosphere to consolidate the laminate.

The vacuum-assisted molding technique creates pressure on a laminate during its curing cycle. The pressurization of the laminate fulfills several functions. First, it removes trapped air between layers. Second, it compacts the reinforcing layers by force transmission, providing more uniform laminates. Third, it prevents the orientation of the laminate from changing during curing. Fourth, it reduces humidity. Fifth, and perhaps the most essential function, the vacuum technique optimizes reinforcement-matrix relationships in composite parts. All of these advantages have been used for years to maximize the physical properties of composite materials. This technique is today the preferred technique for shipbuilders looking for extremely light structures with high structural requirements.

Jute composites impregnated with styrene resins were manufactured using the VARTM technique (Singh et al. 2022). The study compared the VARTM process with the hand lay-up method to determine jute fibers' tensile and flexural strength. The authors concluded that the tensile and flexural strength properties of the samples processed by VARTM were more robust than those processed by hand lay-up.

Untreated jute fiber has affinity for moisture absorption, weakening the bond between the fibers and the matrix. Jute fibers are treated with alkaline NaOH to reduce the moisture content (Sathishkumar et al. 2017). Composites containing four layers of untreated and treated jute fiber mat impregnated in a vinyl ester resin were manufactured using the vacuum bag method. The alkali treatment promoted that the composites presented greater rigidity and resistance than pure resin. The authors observed that the jute fiber treated with a higher concentration of alkali represented a more significant improvement in compatibility with the matrix. However, the excess treatment produces the degradation of fibers with high resin absorption (E. A. Franco-Urquiza, Vázquez, and Escalante-Velázquez 2021). The effect on processing conditions is analyzed in woven jute preforms treated with NaOH (5 wt%) for 24 hours at room temperature (Rodriguez, Stefani, and Vazquez 2007). The composites were prepared using a vacuum infusion process. The treatment was detrimental to the mechanical properties of the fibers. The impregnation times increased in the treated jute due to the increased exposed area and the resistance to flow. The flexural and impact properties of the treated jute composites decreased mainly at the lower mechanical properties of the fibers.

Similar observations were found in other work (van Oosterom et al. 2019), where the resistance is partly affected by the resin's weight. A lower amount of resin in the manufacture of jute composites by VARTM results in an increase in their mechanical properties compared to the composites manufactured by the hand lay-up technique (van Oosterom et al. 2019; Singh et al. 2022). Other authors highlighted that the jute fiber/epoxy composites manufactured by hand lay-up contain a more significant number of defects than those manufactured by VARTM due to the absence of a vacuum (Bakhshi and Hojjati 2020; Elkington et al. 2015). Torres et al. (M. Torres-Arellano et al. 2020) developed a particular work on manufacturing and mechanical characterization of natural-fiber-reinforced biobased epoxy resins. Biolaminates are attractive to various industries because they are low-density, biodegradable, and lightweight materials. The manufacture of the biolaminates was carried out through the vacuum-assisted resin infusion process. The mechanical characterization revealed that the Jute biolaminates present the highest stiffness and strength. The rigid and semirigid biolaminates obtained in this work could drive new applications targeting industries that require lightweight and low-cost sustainable composites.

Other authors (Torres et al. 2020) used the VARTM process to prepare natural fibers reinforced bio-based epoxy resin. ZnO nanoparticles were added to the bio-based epoxy before the impregnation process. Viscoelastic and mechanical properties were evaluated and related to the nature of the fibers and filler content. ZnO particles resulted in effective fillers just at low concentrations and induced different reinforcement mechanisms attributed to the interaction between the nature of fibers and nanoparticles.

Resin Transfer Molding

The resin transfer molding (RTM) process can also be used to produce composite components, such as body parts or frames, in the automotive industry. The fibers are impregnated with thermosetting resin in a closed mold, the main difference from the VARTM process. Closed molds provide a clean surface on both sides of the component or part. In RTM, the pressure is higher than the atmospheric, which increases the quality of the composite component. Once the composite is cured, the component is removed from the mold and further processed.

The RTM process has been used to manufacture composites with various fiber contents, up to 20.6% by volume. The wetting of the fibers is usually excellent. Resin injection time was found to increase with high fiber content due to the low permeability. Keeping the temperature of the mold constant is the key to obtaining a fast and homogeneous cure for the component (Rangappa et al. n.d.).

Sanjay and collaborators (Rangappa et al. n.d.) prepared sustainable composites using the RTM process. This process was used to study the effect of water absorption on the mechanical properties of composites with natural fibers. The water absorption test was carried out by immersing the specimens in a water bath at room temperature for a while. It was found that the water absorption process of these composites approximates the Fickian diffusion behavior. Diffusion coefficients and maximum water uptake values were evaluated; the results showed that both increased with an increase in fiber content. The tensile and flexural properties of specimens immersed in water were evaluated and compared to dry composite specimens. The results show that the swelling of the flax fibers due to water absorption can positively affect the composite material's mechanical properties. This study showed that the RTM process could be used to manufacture natural fiber-reinforced composites with good mechanical properties, even for potential applications in a humid environment.

Safiee et al. (Safiee et al. 2011) studied the properties of pultruded jute fiber-reinforced unsaturated polyester composite (PJFRC). Compression and bending tests were performed to study the mechanical properties, while thermomechanical tests were used to study the thermal properties of the composites. The morphological aspects of the composites were also evaluated using a scanning electron and light microscope. The compressive and bending stress-strain curves showed a linear portion in the initial loading phase, followed by plastic deformation and creep. The storage modulus is highly dependent on temperature and degrades with increasing temperature.

Other authors (Verhagen et al. 2009) have detailed the feasibility of replacing glass fibers and carbon fibers with vegetable jute fibers in polymeric composites. The work carried out focuses on the importance of having a comparison of the mechanical properties of these natural fiber composites with those of traditional fiberglass and carbon composites. Glass and carbon fiber composites are currently used in many applications that may not require high-strength materials, a lower strength jute fiber composite may be suitable. Natural fiber composites are currently used primarily for non-structural applications. This research aims to determine if these natural fiber composites have the mechanical properties that allow them to be used in structural applications. Fiberglass and jute fiber composites were manufactured through RTM and fiber layout after equalizing both fiber volume and mass fraction. Specimens were tested to determine the ultimate tensile strength, the ultimate load before failure, the tensile modulus of elasticity, and the tensile strain. The fiberglass specimens could withstand tensile loads about five times higher than those of the jute fiber samples. The fiberglass specimens were also+ stiffer than the jute fiber parts. The jute specimens' performance after matching the fiber mass fraction (with the fiberglass piece) was generally better than the samples made after matching the fiber volume fraction.

It is increasingly common to find research dedicated to substituting synthetic fibers to develop green composites and place them in various industrial sectors. One of these works uses woven jute fibers to investigate water absorption and the effect of squatting in a humid environment (Masoodi and Pillai 2012). For this, synthetic and biobased epoxy resins were used. Several of these composite samples were made using a low-pressure resin injection process, similar to resin transfer molding, and consisted of three compositions: pure resin, pure resin with a single layer of jute fabric, and pure resin with two layers of jute fabric. It was observed that the rate of moisture diffusion in composite materials increases with the jute fiber-to-epoxy ratio. The type of epoxy used as a matrix seemed to influence the percentages of moisture absorption of the composites. The study also showed that water absorption and swelling were higher in the biobased epoxy than in the epoxy composites. Investigations of composites prepared using RTM technology continue, and a higher incidence is expected in the following years.

Compression

Compression molding is a manufacturing process in which a measured amount of molding material, usually preheated, is compressed to the desired shape using two heated molds. Compression molding is a production method that requires relatively low pressure for molding compared to other manufacturing techniques such as injection. This process also wastes little material, which is advantageous when working with expensive materials. Compression molding is often preferred for low- to medium-volume production of parts and is often the most cost-effective manufacturing method if simple, large, primarily flat parts need to be produced. Some designs' curves, nooks, and crannies are acceptable, but extreme angles and deep depressions can be challenging to achieve with compression molding. Thanks to the lower pressure involved in the process, tooling costs are affordable, and molds typically last long without warping or needing replacement. To offset the cost associated with the long cycle times of compression molding, manufacturers can use a multi-cavity mold to produce multiple parts in the same cycle. The structural stability of compression molded parts is very high. Although compression molding offers many advantages, it also has its limitations. The labor cost associated with compression molding can be relatively high, because long cycle times translate into more labor hours.

There are several variants of compression molding: Sheet Molding Composite (SMC), Bulk Molding Composite (BMC), and Glass Mat Thermoplastics (GMT) stand out. Thermoplastic composites use the latter, while the SMC and BMC processes are primarily used for composites with fibers preimpregnated thermosetting resins. These processes offer short cycle times, a high degree of productivity, and automation with dimensional stability, which is why they are used in various applications in the automotive industry (Chabros et al. 2020; Huda et al. 2008; Robson and Goodhead 2015; Suddell and Evans 2010; Taylor et al. 2002). Some jute fiber-reinforced epoxy composites are manufactured using manual lamination followed by a compression molding technique (Katayama et al. 2006; Selvan 2021; Singh et al. 2022; Singh and Zafar 2020). Bapan et al. (Adak and Mukhopadhyay 2016) developed jute fiber composites using conventional compression molding and hand lay-up methods. Five layers of alkali-treated jute fabrics were impregnated with ionic liquid. The impregnated layers were stacked symmetrically, maintaining the warp and weft directions of the fabric. The curing process was performed in a compression molding machine to prepare compact composite laminates. Light microscopy and SEM observe the jute fiber composite laminates' surface morphology and cross-section. It was found that the interlayer bonding morphology and mechanical properties are highly dependent on pressure, temperature, and dissolution time.

Some works developed jute fiber reinforced epoxy composites by changing the number of jute fiber layers and adding various filler contents of zinc oxide (Rajasekhar, Ganesan, and Senthilkumar 2014; Sathishkumar et al. 2022). Compression molding prepared jute composites and tested for tensile, flexural, and impact properties (Sathishkumar et al. 2022). The results reveal that increasing the number of layers of jute fiber increased the mechanical properties. The maximum resistance was obtained in the double-layer composites. By incorporating zinc oxide filler, the filler content gradually increased the mechanical strengths, which increased the bonding effect between the fiber and the matrix. The higher content of zinc oxide filler decreased the mechanical properties due to lower resin content and debonded layers.

Thermoplastic Composites

Due to their structural advantages, thermosetting resin composites have great industrial potential. On the other hand, these composites have disadvantages such as the high cost of raw materials and the costs associated with manufacturing processes, including the curing time during the manufacture of a component. Thermoplastic resins show great potential to produce strong, lightweight composites that are easily recyclable. Thermoplastic composites use pulverized fibers in the form of particles or short fibers as reinforcement. In most applications, natural fibers are used to reduce weight without losing structural stability (Barkoula et al. 2008; Chauhan et al. 2019; Katayama et al. 2006; Khan et al. 2012; Mathur 2021; Memon and Nakai 2013; Muzammil et al. 2022; Pailoor, H. N. Narasimha Murthy, and Sreenivasa 2021; Shahinur et al. 2022; Velasco-Parra, Ramón-Valencia, and Mora-Espinosa 2021).

Thermoplastic composites offer multiple advantages, including concise production times and low total costs, compared to thermoset composites. A typical thermoplastic part can reduce manufacturing time by 20-30% compared to thermoset composites (Chandekar et al. 2020; Chauhan et al. 2019).

Currently, the automotive sector develops internal components for car doors using thermoplastic composites, reducing the weight of doors by half (Chauhan et al. 2019; Velasco-Parra et al. 2021). Additionally, these components can be welded or overmolded to produce advanced geometries. Therefore, thermoplastic composites also have the advantage of being flexible in their design and adopting new forms without losing their physical properties.

Thermoplastic composites are mixed using extrusion technology. Sometimes the mixing is carried out in injection molding, although it is clear that this technology is more efficient for molding a piece than mixing materials. Once the thermoplastic composites have been created in powder or pellets, they are used as raw material for their transformation into products through various processes such as injection, extrusion, thermoforming, blowing, calendering, and compression, among others.

Extrusion Process

The thermoplastic composite, pellets or powder, is poured into a feeding hopper and is pushed by an endless screw (located inside a cylinder with the right temperature for the granules to become liquid) towards the nozzle (Chandekar et al. 2020; Chestee et al. 2017). This nozzle comes out with the shape of the profile to be manufactured and is cooled. Extrusion is a continuous manufacturing process, so at the end of the extrusion line, the material is cut into pellets for subsequent transformation, rolled using a winder, or cut the profile obtained.

The industrial applications of thermoplastic composites are of relevant interest today. Sanvezzo and collaborators (Sanvezzo and Branciforti 2021) approached the recycling of industrial waste of jute fibers to develop Jute/PP composites mixed with a compatibilizer and calcium nano carbonated. The components were mixed in a co-rotating twin screw extruder, and dumbbell-shaped specimens were produced by injection molding. A masterbatch was previously processed to improve the dispersion of the nano carbonated in the polymer matrix. The composites were exposed to accelerated aging, tested by contact angle and tensile measurements, and evaluated for visual appearance before and after accelerated aging. The morphological evaluation was carried out through SEM observations. The processing techniques provided composites with 50 wt.% of waste, good dispersion of the nanoreinforcement, no thermal or mechanical degradation of the fibers, and better mechanical properties than pure PP. The presence of industrial residues led to a 135% improvement in the elastic modulus of the PP matrix and decreased strain at break from 435% to 5%. The nanoreinforcement acted as mechanical reinforcement and decreased wettability. Masterbatch and direct extrusion increased the stiffness up to 18% and 16% respectively.

The composite processed after the manufacture of the masterbatch and without compatibilizer showed a 186% improvement in the elastic modulus of the matrix, the lowest wettability, and the best average performance under accelerated aging. Accelerated aging promotes a slight reduction in stiffness but a drastic drop in elongation of materials. The authors highlighted the potential use of these composites in various commercial applications due to their properties and processing flexibility, which is relevant both from a sustainable and economic point of view.

Other authors prepared partially biodegradable jute/PP composites in extrusion followed by compression molding (Huq et al. 2013). Short jute fibers acted as potential reinforcing agents for the PP matrix. The composites were exposed to gamma radiation. Irradiated composites gained significantly higher strength and modulus than non-irradiated composites. Six months of soil degradation tests revealed that the irradiated composites retained more mechanical properties than their control counterparts. A reaction mechanism between jute and PP due to irradiation treatment was proposed.

Some research addresses the effect of processing on the aging and rheology of thermoplastic composites reinforced with jute fibers (van den Oever and Snijder 2008). The PP pellets comprise up to 50 wt. % of jute fiber. Rheological analysis shows that the viscosity of the composites follows the same shear rate dependence as PP and is at the same level as glass-PP composites. Mechanical properties show minimal variation and exhibit strength and stiffness values in the upper range of competitive injection molding natural fiber-reinforced composites (Chandekar et al. 2020; Franco-Urquiza 2022; van den Oever and Snijder 2008; Velasco-Parra et al. 2021). Mechanical performance gradually reduces with prolonged thermal load and immersion in water (Xian, Guo, and Li 2022). The low water diffusion coefficient indicates that the fibers do not form a continuous web throughout the polymer.

Jute/PP composites are prepared in an extruder, followed by compression molding (Roy et al. 2012). The mechanical properties of the composite materials increased up to maximum fiber content, then decreased. Through the same process, composites with variable percentages of sodium bicarbonate were manufactured at a constant content of jute. The composites' mechanical properties, soil degradation tests, and water absorption capacity were adequately investigated (Huq et al. 2013; Roy et al. 2012). In addition, the density of the composite was reduced (Roy et al. 2012).

Polypropylene/jute fiber composites have been studied using impact modifiers and maleated polypropylene (MAPP) as compatibilizers. Composite materials were manufactured by extrusion and compression molding. After the extrusion process, the effect of fiber length on the mechanical performance and toughness of the composites was investigated. The authors indicated that incorporating compatibilizers improved the composites' energy absorption (Ranganathan et al. 2015). The highest impact strength was found with the addition of 10% by weight of the impact modifier, but the higher concentration of the impact modifier negatively affected the tensile and flex properties.

Through the literature consulted, it can be seen that most of the research carried out on jute fiber thermoplastic composites employs a polypropylene matrix, which is attributed to the low cost of this polyolefin, its extensive knowledge of processability, and the relatively easy acquisition for the development of composites reinforced with this natural fiber. The previous is the beginning to make room for more daring research, using other types of polyolefin such as polyethylene and biodegradable resins such as lactic acid.

The twin screw extrusion processes report better processing for thermoplastic composites, although it also promotes the degradation of the natural reinforcement. Grande et al. (Grande and Torres 2005) studied the mechanisms of organization and degradation of natural fibers caused by single screw extrusion processing. They used high-density polyethylene (HDPE) as a matrix. Jute and sisal were employed as reinforcements with initial lengths ranging from 5 to 10 mm. Fiber distribution and bubble formation were related to the mechanical properties of these composites, particularly tensile strength. The higher processing temperatures, the fibers show greater alignment in the flow direction.

Other research articles evaluated the effect of surface treatment of jute fiber on the properties of biodegradable PLA composites (Rajesh and Prasad 2014; Shahinur et al. 2022). PLA composites with untreated, NaOH-treated, and (NaOH + silane)-treated jute were prepared in a twin-screw extruder. The effects of jute fiber surface treatment were examined at the fiber-matrix interface by FT-IR spectroscopy, which confirmed the formation of hydrogen bonds and covalent bonds between PLA and jute fibers. Polarized light optical microscopy (PLOM) images showed enhanced trans-crystallinity at the matrix-fiber interface, while SEM showed matrix-covered fiber surfaces in the case of treated jute fiber-reinforced composites. The addition of untreated jute fibers in PLA increased tensile strength and modulus, while composites reinforced with surface-treated jute fibers exhibited higher mechanical properties (Zafar, Maiti, and Ghosh 2016).

Injection Molding Process

The plastic injection technique consists of manufacturing a piece inside a mold. The polymers are injected under pressure into the mold through a nozzle, and the pressure is kept constant while the piece cools, giving it its final shape. Once the process is finished, we will have the final piece obtained just by opening the mold. An extensive literature on plastic injection can be consulted (Zafar et al. 2016).

For the past two decades, researchers have been trying to explore eco-friendly materials that would significantly reduce reliance on synthetic fibers and their composites (Rabbi, Islam, and Islam 2021). As stated before, natural fiber-based composites possess several excellent properties (Hashemi 2002). They are biodegradable, non-abrasive, low cost, and have a lower density, which has led to a growing interest in using these materials in industrial applications (Rabbi et al. 2021). Before manufacturing, the fiber and the matrix must be combined. Many researchers have adopted different mixing techniques. Single or twin-screw extruders, two-roll mill ball machines, and K-mixer are common types used for mixing short fiber and polymer. Among various composite manufacturing processes, injection molding is famous for mass production. Thermoplastics produced by the injection molding technique have been used for many decades, from small household items to extreme-performance automobile parts (Bex et al. 2019; Jeong et al. 2022; Ogorodnyk and Martinsen 2018; Rubin 1991).

As previously mentioned, thermoplastic composites are usually first developed in a single or twin screw extruder and subsequently molded into a composite part. Rabbi et al. (Rabbi et al. 2021) prepared jute/PP composites using a twin-screw extruder and injection molded specimens. The effects of staple/continuous fibers, coupling agent, and fiber ratio on mechanical properties were investigated. The tensile and flexural moduli of continuous jute/PP were higher than those of staple fiber/PP. The tensile, flexural, and impact strengths were higher in staple fiber/PP and the elongation at break. The coupling agent improved tensile and flexural strengths, increasing fiber content, while impact strength and elongation at break decreased with fiber loading.

Rana et al. (Hashemi 2002) made a jute fiber polypropylene composite to observe the influence of fiber filler, impact modifier, and compatibilizer in the composite. A high shear K-mixer was used for better mixing, and dry pellets were prepared before injection.

A complete review article gives an extensive bibliographic on bio-composites fabricated by the injection molding method (Rabbi et al. 2021).

Some studies evaluated the effects of optimization of materials and composition processes on the properties of natural fiber composites (Sun, Han, and Dai 2009). The thermal stabilities of sisal and jute fiber were compared by thermogravimetric analysis. The influences of fiber content, coupling agent, fiber geometry, and fiber distribution on properties were also investigated. Sisal fiber was found to have more thermal stability than jute fiber. Adding coupling agents, long fiber length, and uniform fiber distribution led to higher performing composites.

The mechanical properties of polylactic acid (PLA)-based green composites with jute fibers have been extensively investigated (Arao et al. 2015). Short fiber composites exhibited optimal mechanical performance. Although the extrusion process lowers the residual fibers' overall aspect ratio, it facilitates the jute reinforcement's dispersion and prevents its decohesion from the matrix. Jute/PLA composites transfer the load efficiently from the matrix to the fiber and improve interfacial strength. The jute/PLA composites also illustrated high mechanical performance (Bledzki and Jaszkiewicz 2010). The adequate process condition for PLA-based green composites via twin-screw extrusion with lower temperature profiles exhibited properties superior to those produced at higher temperature, due to reduced thermal degradation of the natural fiber during composition (Gunning et al. 2013).

Jute-reinforced PP and PLA composites were manufactured using the direct fiber feed injection molding (DFFIM) process (Sarasook et al. 2020). Jute fibers were fed directly into the barrel of the molding process to eliminate fiber breakage during the extrusion process. The tensile strength of composites decreased, and the modulus increased compared to pure PP. Using maleic anhydride grafted polypropylene (MaPP) can improve the interfacial bonding between jute fiber and PP matrix as observed by SEM, resulting in increased tensile strength (Rabbi et al. 2021; Sanvezzo and Branciforti 2021; Sarasook et al. 2020). Therefore, in the case of jute/PLA composites, the surface of the jute fibers is treated with sodium hydroxide (NaOH) and a silane coupling agent to improve interfacial adhesion (Rangappa et al. n.d.; Sarasook et al. 2020). In addition, the tensile properties of NaOH-treated jute/PLA and NaOH+silane-treated jute/PLA composites were improved compared to untreated composites (Zafar et al. 2018).

Compression Molding

Compression molding is used to manufacture both thermoplastic and thermosetting composite parts. Compression molding can be used to make polymer composites with long, short, and intermediate-length fibers. Compression molding is preferred for mass production of composites due to high precision and low processing times.

Jute fabrics and composites based on PP reinforced with unidirectional jute fiber and linear low-density polyethylene (LLDPE) were successfully prepared by compression molding technique (Niloy Rahaman et al. 2019). Jute fiber-based composites showed higher mechanical properties than jute-based fabrics. The polypropylene-based composites showed better mechanical properties than those of LLDPE.

Another study presents improvements in the mechanical performance of jute fiber composites (Hasan et al. 2021). Unidirectional jute composite laminates comprising stitched and sized preforms were fabricated using a compression molding technique. Jute composites exhibited significant mechanical properties, with the most remarkable improvement observed in the case of the PVA-sized alkali-treated sample, thanks to the excellent compatibility between the alkali-treated and sized jute fibers. Another advantage of thermoplastic composites is that they can be drilled, compared to thermosetting composites which are damaged during the drilling process.

Additionally, some studies confirm that natural fiber composites are more suitable for drilling than synthetic fiber composites (Debnath, Singh, and Dvivedi 2014).

Hybrid Technology for the Manufacture of Jute Composites

The different manufacturing technologies for FRP composites have specific characteristics depending on the part's production requirements. Historically, the shaping of FRP composites has been manual and limited work. However, these processes have given way to others in search of automation and repeatability in today's various manufacturing processes. The prepreg material used in FRP is the reinforcing fiber in the form of fabric or unidirectional fibers previously pre-impregnated with the polymer resin. In this way, it is possible to provide the right and necessary amount of resin, optimizing the mechanical properties of the materials and the weight of the pieces to the maximum. The process of manufacturing parts using prepreg begins with cutting the material into the necessary shapes. After that, a specialized operator will laminate the material. He will place the cuts previously made on the mold, following its geometry. This process is vital since the thickness uniformity, the material's overlaps, and the perfect copying of the surfaces must be controlled. Thus, prepreg shows one of its most significant advantages, making it possible to create complicated shapes relatively simply and cleanly. Finally, the part will be processed by curing the thermosetting matrix. The most common techniques for curing prepregs are vacuum bagging and autoclave. The differences between these processes are the curing temperature and especially the pressure. The higher the latter, the better mechanical properties are obtained since the compaction of the material increases, and the appearance of air bubbles is avoided to a greater extent. However, there is a limit where the material is choked, and the resin cannot flow. On the other hand, with the temperature, it is possible to control the curing times, another parameter of vital importance. It is essential to know and have control of the temperature and pressure slopes and curing times of the materials to carry out the process satisfactorily. In an autoclave, temperatures range from 90 to 200°C with up to 10 bar pressures.

Because jute fiber is a hydrophilic natural fiber, the development of prepregs is counterproductive to the development of natural fiber composites. Therefore, composites are manufactured through hybrid technologies to provide better properties of the composite.

The work carried out by Alim et al. (Alim et al. 2022) evaluated the effect of superheated steam (SHS), jute fiber, and PLA, which were synthesized by the melt mixing method. The objective of this treatment was to enhance the fiber-polymer interfacial bond. The action was carried out in a superheated steam oven at various times and temperatures. The biocomposites were evaluated in terms of mechanical characteristics, dimensional stability, and morphological properties. The results showed that the treatment offered the best tensile characteristics. Due to the presence of SHS-Jute, the tensile, impact, flexural and dimensional stability of the biocomposites have been improved. The interfacial bond improves the flexural strength of the SHS-Jute-PLA biocomposites. The SHS treatment reduces the content of hemicellulose in the jute fiber, which causes a reduction in water absorption. SHS-Jute-PLA biocomposite appeared with promising characteristics for use as a green and eco-friendly substitute particle board material.

Several works detail the use of nanoparticles to reinforce natural fiber composites. Hossen et al. (Hossen et al. 2020) fabricated jute/PLA composites reinforced with Montmorillonite clay. PLA/nanoclay pellets were obtained in a twin screw extruder. The nanoclay modifies the melting temperature of the polymers because of the moisture retention of the organo-modifiers (Franco-Urquiza 2021; Franco-Urquiza et al. 2010; Maspoch et al. 2009). PLA/nanoclay and jute fibers were combined, and PLA reinforced with jute/nanoclay fiber was manufactured by the autoclave molding method (Hossen et al. 2020). The effect of the vacuum condition in the manufacturing process on the mechanical properties of PLA reinforced with jute fiber/nanoclay was investigated. The vacuum condition in the heating process prevented thermal oxidation of the jute fiber, resulting in high tensile strength even after heating in manufacturing. The tensile property of PLA was improved by adding jute fiber and nanoclay. The nanoclay composite increased Young's modulus and tensile strength, although it decreased breaking stress (Franco-Urquiza et al. 2015). Adding jute fiber in PLA/nanoclay reduced breaking stress by increasing Young's modulus and tensile strength. On the other hand, a lower amount of nanoclay was enough to improve the Young's modulus of PLA than jute fiber. The combination of nanoclay and jute fiber was effective as reinforcement for PLA.

Two stages procedure combining twin-screw extrusion and compression molding implies higher cost and degradation of mechanical properties (Debnath et al. 2014). On the other hand, in-line compression molding, also called direct molding of long-fiber thermoplastics (DLFT), is a single-stage process in which the fibers are introduced directly into the molten polymer in the extruder, and extrusion heat is instantly removed during molding, resulting in reduced cycle times (Pailoor, H N Narasimha Murthy, and Sreenivasa 2021).

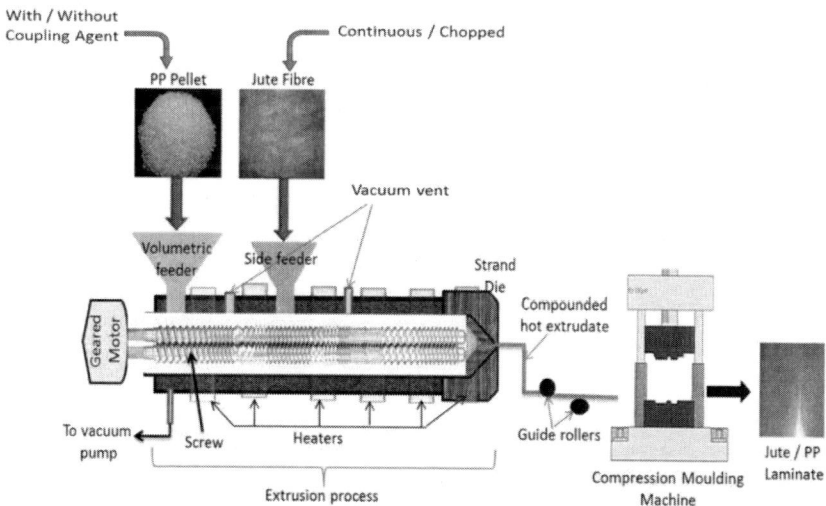

Figure 3. Schematic representation of Jute/PP in-line compression molding (Pailoor, H N Narasimha Murthy, et al. 2021).

In-line compression molding is gaining importance for producing structural and semi-structural components for automotive applications. It reduces cycle time through a one-step process where the fibers are introduced directly into the polymer melt in the extruder, which is directly molded, as schematized in Figure 3.

Adding natural fibers used as reinforcement dramatically appeals to the construction materials industry as natural fibers are cheaper, biodegradable, and readily available. The feasibility of using jute fibers as reinforcement in mortars was analyzed, taking advantage of the mechanical properties of resistance to compression and bending (da Fonseca, Rocha, and Cheriaf 2021). The mortars were reinforced with raw and treated fibers. Fiber-free mortars were used as a reference. The treated fibers improved the flexural strength of the mortars and prevented the samples from snapping, in contrast to the brittle behavior of the reference samples.

Another study explored the efficacy of woven jute/polyester (JP) composite tubes as energy-absorbing structural countermeasures (da Fonseca et al. 2021). In this sense, the behavior of 3- and 4-layer JP composite tubes with square sections and double hats subjected to quasistatic axial and impact loads are considered and compared with 4-layer glass-polyester (GP) tubes. The JP pipes can withstand impact loads in axial and bending modes.

Hybrid Materials

The popularity of jute fiber composites and hybrids is primarily due to increased environmental concerns and pollution. Hybrid jute fiber composites have been widely used in various applications. Hybrid synthetic fiber composites have better mechanical properties than natural fiber composites. Synthetic fibers have better fiber/matrix adhesion, which increases the overall mechanical properties of the composite material. However, the high cost of making synthetic fibers has limited the use of these hybrid composites in various applications (Ashraf et al. 2019). Some issues related to synthetic fiber hybrid composites are environmental concerns, recyclability, biodegradability, and reuse (Yang et al. 2011). Researchers seek to improve the properties of natural fiber composites with the inclusion of synthetic fibers. The hybridization of natural fibers with glass fibers significantly improves the mechanical properties of the composite (Yang et al. 2011).

The mechanical properties of hybrid glass/jute fiber reinforced polyester composites were evaluated. Young's modulus increased with increasing glass fiber content, while Poisson's ratio decreased because of higher transverse strain and lower longitudinal strain in the jute fiber composite compared to the jute/glass fiber composite (Ahmed, Vijayarangan, and Naidu 2007). Moisture content decreased mechanical properties such as tensile strength and Young's modulus because moisture disrupts fiber/matrix adhesion (Aquino et al. 2007). The presence of jute fiber in a composite with a stacking arrangement combining (Ashraf et al. 2019; Ramprasath et al. 2020; Selver, Ucar, and Gulmez 2018). The Jute/E-Glass reinforced hybrid composite was manufactured using the hand lay-up technique. Five different hybrid combinations of jute and E-glass were tested. The jute/E-glass hybridization imparts better mechanical and wears properties than the single fiber reinforced composite (Jha, Samantaray, and Tamrakar 2018). The combinations indicate that although jute-based composites exhibit relatively inferior mechanical properties, their properties are improved by multiples when combined with E-glass fiber. It further indicates that the incorporation of E-glass fiber results in an improvement in the erosion wears resistance of the jute/E-glass fiber composites.

The hybridization of natural and synthetic fibers leads to the optimal mechanical properties of the composites. The effect of stacking sequence on PBS-based glass-jute hybrid (GJ) composites were evaluated (Ghani et al.

2022). The composites were manufactured using the compression molding method. The stacking sequence of the fiber layers significantly affects the overall performance of the GJ hybrid composites. The composites with fiberglass layers on their outer surfaces showed optimal mechanical, thermal, and water resistance properties (Zamri et al. 2011).

The demand for fiber-reinforced composite materials is increasing in structural applications due to their crucial characteristics, such as stiffness, strength and durability, and low-cost processing benefits (Sathiyamoorthy and Senthilkumar 2020). Stacking sequence in hybrid jute/carbon composite laminates on the tensile, impact, micro-hardness, water absorption, and thermal behavior were studied. The manual layout was used to fabricate the composite laminates with four stacking sequences. The experimental results showed that the hybridization process improved the composites reinforced with jute properties. FT-IR and XRD analyses revealed that the alkalinization process removed binding components such as lignin and hemicelluloses from the raw jute fiber, resulting in a higher crystallinity index. Hybrid composite with jute/carbon/carbon/jute stacking patterns has the highest tensile strength compared to other stacking sequences. The carbon/jute/jute/carbon fabric stacking sequence exhibited higher impact strength and better moisture resistance. Incorporating jute with carbon decreased tensile strength and impact strength compared to carbon-reinforced composites.

The flexural behavior of hybrid composite becomes essential when deciding the fiber stacking sequence. Jute, glass, and carbon fibers are used as polyester resin reinforcement manufactured using the vacuum bagging method (Murdani and Amrullah 2021). The result shows that the fiber stacking sequence gives significantly different bending behavior regardless of the effect of fiber strength.

Hybrid composites reinforced with jute and kevlar with different stacking sequences, namely jute-kevlar-jute (JKJ) and kevlar-jute-kevlar (KJK), using the vacuum bagging method and compared with the jute-jute-jute (JJJ) and kevlar-kevlar-kevlar (KKK) composites (Mahesh, Mahesh, and Harursampath 2021). The physical characterization revealed no appreciable variation in the density of the four proposed composites. Hybridization results in better water absorption resistance than the total natural fiber reinforced composite. Mechanical characterization reveals that JKJ composite can be a potential replacement for KKK composite in structural applications that require good tensile and flexural strength. The KKK composites showed better impact resistance. However, hybridizing the composites results in higher impact resistance than the total natural fiber reinforced composite.

Hybridization with Natural Fibers

Composites combining two or more natural fibers in a polymeric matrix are known as hybridization with natural fibers (Ashraf et al. 2019). Boopalan et al. (Boopalan, Niranjanaa, and Umapathy 2013) studied jute/banana fiber reinforced epoxy composites' thermal, mechanical, and water absorption properties. Flexural, tensile, and impact strengths were highest for an epoxy hybrid composite's 50/50 weight ratio of jute and banana fibers. Water absorption was found to be dependent on fiber content and stacking sequence. When the hybrid composites were subjected to moisture, a significant reduction in flexural and tensile properties was observed due to water absorption (Akil et al. 2014). The stacking sequence and sodium bicarbonate treatment on the mechanical properties of a hybrid jute/linen epoxy composite were studied (Fiore and Calabrese 2019). Sodium bicarbonate treatment significantly improved the hybrid composite's quasi-static properties and helped improve the flax-based composite's mechanical properties, while the jute-based composite showed a slight decrease in mechanical properties (Fiore and Calabrese 2019).

Jawaid et al. (Jawaid et al. 2011) studied the chemical resistance, void content, and tensile properties of a three-layer epoxy-reinforced jute/oil palm fiber hybrid composite. The hybrid composite had a slightly higher chemical resistance than the pure composite. When oil palm fiber was used as the skin layer in a three-layer composite, it had a higher void content due to fiber/matrix compatibility issues. The jute/oil palm fiber/jute three-layer composite had a higher tensile strength compared to when the jute fiber layer was sandwiched between oil palm fiber layers because jute fibers are more compatible with epoxy resin. In subsequent research (Jawaid, Abdul Khalil, and Abu Bakar 2010), the hybrid composite increased flexural strength and modulus compared to the pure palm fiber composite. Alkali-treated jute-palmyra hybrid fibers improve the mechanical properties as compared to the hybridization of synthetic fibers such as glass fibers (Shanmugam and Thiruchitrambalam 2013).

Singh and collaborators (Singh and Zafar 2022) prepared multilayer composites of hybrid woven fiber mats using a microwave-assisted compression molding process. The hybrid composite consisted of jute (J) and kenaf (K) fiber as reinforcement, while HDPE acted as a matrix. The stacking sequence resulted in different hybrid composites: JJJ, KKK, JKJ, and KJK. JKJ composite exhibited a maximum flexural strength, while KJK exhibited a maximum flexural modulus. The tensile strength, impact strength, and Shore D hardness were maximum in the case of the KKK composite because the stacking sequence has a nominal effect on the flexural strength, tensile strength, and impact strength of composite materials.

Another study looked at the effects of kenaf and jute fiber stacking sequence on the tensile and flexural properties of kenaf/jute hybrid composites. The authors prepared Kenaf/jute/kenaf (K/J/K), jute/kenaf/jute (J/K/J), and pure epoxy (EP) composites using the manual lamination technique (Khan et al. 2021). The K/J/K hybrid composites exhibited higher tensile strength and flexural strength than the J/K/J hybrid composites. Hybrid composites K/J/K also show better tensile and flexural modulus than J/K/J hybrid composites and pure epoxy composites.

The effects of different fiber bundle loadings and the modification of bagasse fiber surfaces in hybrid fiber-reinforced epoxy composites have been studied (Saw and Datta 2009). Fiber surface modification reduced the hydrophilicity of the fiber bundles, and significantly higher mechanical properties of the hybrid composites were observed with the SEM. SEM analysis of the fiber and fractured surfaces of the composites showed better compatibility at the interface between the chemically modified fiber bundles and the epoxy resin.

Other Structures

The use of sandwich composites has been increasing daily in structural applications due to their high strength-to-weight ratio, high environmental resistance, and good thermal insulation properties. Jute fiber is an alternative to conventional materials to combine properties with Kevlar fibers as reinforcement in epoxy by a manual lamination process. The cardboard honeycomb structure was used as the core of the sandwich structures. The results reveal that hybrid sandwich composites have a superior tensile, impact, and flexural strength compared to sandwich composites. The compressive strength remained unchanged in all cases.

Rajole et al. (Rajole, Ravishankar, and Kulkarni 2020) explored the use of jute/epoxy (JE) and jute/rubber/epoxy (JRE) natural fiber sandwich composites for ballistic energy absorption. The results were evaluated analytically and by finite element analysis (FEA), as schematized in Figure 4. The analytical results are found to agree well with the results of the FEA, with a maximum error of 9%. The thickness influences energy absorption. The experimental and FEA study showed that JRE sandwiches have better energy absorption than JE plates. The energy absorption of a JRE sandwich is approximately 71% higher than that of JE plates. Damages obtained from FEA and tests are in good agreement. SEM analysis confirms that the composites failed due to fiber breakage and fragmentation.

Alpha fiber-based core and a hybrid polymeric matrix composite (jute and metal mesh) as skin were evaluated by Laraba et al. (Laraba et al. 2022). The resulting sandwich yield was higher than other bio-based sandwiches, such as cork-based sandwiches, but with a higher density. Hybridization of jute improves the stiffness of the hide but reduces the strength. The skin stacking sequence strongly influenced sandwich rupture, and the presence of metal mesh at the core/skin interface led to delamination, reducing the mechanical properties of the sandwich. Generally, this sandwich could be helpful as a non-structural component in building materials.

The effect of surface layer reinforcement with unidirectional jute/glass fibers and epoxy resin on flexural properties was investigated by Mohan and collaborators (Mohan and Kishore 1985). A substantial increase in flexural modulus and strength was achieved with small amounts of reinforcement. Furthermore, the effect of moisture absorption on flexural properties revealed a significant improvement in moisture resistance of jute-reinforced epoxy with hybridization. Although density analyzes of these hybrids indicate an increase in density with hybridization, the considerable improvement in normalized properties of jute-glass hybrids makes them candidate materials for cost-effective applications.

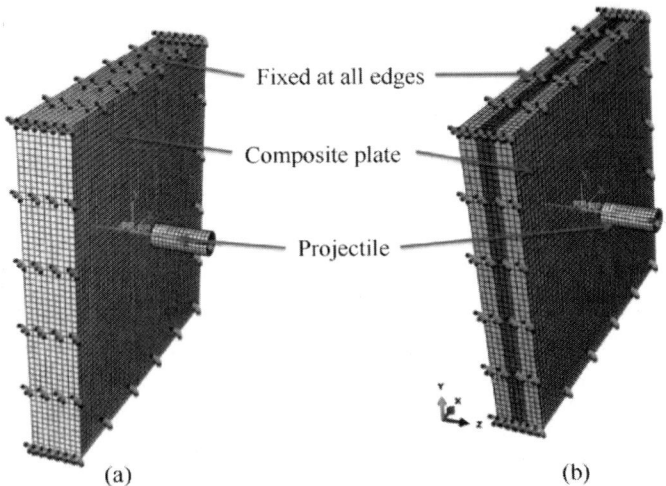

Figure 4. Schematic representation corresponding to the FEA: (a) JE and (b) JRE sandwich composite with boundary conditions imposed (Rajole et al. 2020).

Jute and glass fabrics-epoxy composites were prepared using resin infusion under the flexible tools method (Pandita et al. 2013). The water absorption test was performed on jute woven composites and composite sandwiches. It shows that the thin layers of glass woven composites in the composite sandwich slow down water penetration into the jute woven composites, which are the core materials. The water absorption process applied to the woven composites of jute and its sandwich was modeled using Fick's second law. Glass fabric composites on the outer surface of the sandwich can act as solid facings. The flexural and impact drop weight properties of jute-glass woven composites are higher than those of jute woven composites.

3D Printing of Jute Fiber Composites

3D printing is one of the fundamental elements of industry 4.0, or the fourth industrial revolution, which consists of the automation and digitization of industry. Industrial revolutions are defined by the characteristics of the different emerging technologies of each moment. These characteristics and the new technologies rapidly modify the forms of industrial production while producing a change at an economic and social level. 3D printing is one of the leading emerging technologies of industry 4.0. The use and implementation of additive manufacturing, in combination with other technologies, is producing an evolution in the industry towards intelligent production. 3D printing plays a fundamental role as it is a technology capable of turning a 3D design into a product. 3D printing increases design flexibility and allows the manufacturing of all personalized objects, mainly prototypes, without needing expensive molds and manufacturing tools. 3D printing has reached technological maturity very quickly, and today it is used in medical implants and bioprinting for research in the medical field. Actually, at the Center for Engineering and Industrial Development (CIDESI), 3D printing allowed the rapid manufacture of mechanical ventilators to attend to the health emergency caused by the COVID-19 virus (Figure 5).

3D printing has provided a novel approach for the growth and promotion of natural fiber-based composite materials (Devarajan et al. 2022). Natural fiber-reinforced discontinuous polymers show moderate mechanical properties compared to composites made by conventional processes due to factors specific to the 3D printing process, such as high porosity, limited fiber content, and low fiber aspect ratio (L/ d) (Bi and Huang 2022; Le Duigou et al. 2020; Valino et al. 2019). The next step in the evolution of 3D printing technologies involving natural fibers will be the development of innovative materials for use in 4D printing of shape-changing mechanisms (Le Duigou et al. 2020).

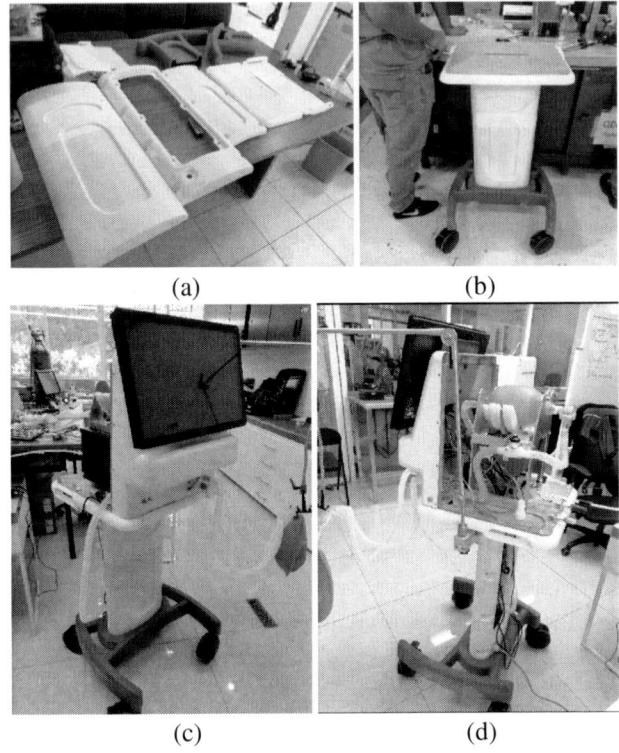

Figure 5. Photographs corresponding to the 3D manufacturing of mechanical ventilators to attend the COVID-19 pandemic: (a) 3D printed samples, (b) assembly of ventilator base, (c) front view of ventilator, (d) system (back view).

Torrado et al. (Torrado et al. 2015) demonstrated the applicability of jute fiber as reinforcement for thermoplastic composite filaments for FDM 3D printing. The thermal degradation of jute fibers leads to voids forming, ultimately weakening the material. The maximum tensile strength of the ABS/jute fiber composite was much lower than the printed ABS.

3D printing of continuous fiber-reinforced thermoplastics based on fused deposition modeling was studied (Matsuzaki et al. 2016). The authors used thermoplastic filament and continuous fibers separately for the 3D printer. The fibers were impregnated with the filament inside the heated nozzle of the printer immediately before printing. PLA was used as the matrix, while carbon fibers, or twisted strands of natural jute fibers, were used as reinforcements.

Unidirectional carbon fiber reinforced thermoplastics showed mechanical properties superior to jute reinforced and unreinforced thermoplastics.

3D printed jute fiber reinforced PLA was prepared and evaluated (Matsuzaki et al. 2016). The authors found that the twisted yarn during impregnation in the die affects the polymer matrix's fiber orientation during the material's printing extrusion. The mechanical properties vary and decrease due to fiber misalignment, poor adhesion between fiber and polymer matrix, and void formation.

Franco-Urquiza et al. (Edgar Adrián Franco-Urquiza, Escamilla, and Alcántara Llanas 2021) evaluated the feasibility of composites using jute fiber fabrics. For characterization, PLA-fused filament was successfully deposited on jute fabrics to print dog bone tensile specimens. The jute fabrics were chemically modified, treated with flame retardant additives, and sprayed with spray adhesive to improve the mechanical properties of the PLA/jute fabric composites. The elastic modulus and strength of PLA were higher than PLA composites, and the plastic deformation of PLA composites was slightly lower than that of PLA. Tomography scans revealed that the fabrics were well oriented and that there was some adhesion between the jute fabrics and the PLA. The viscoelastic properties of the PLA composites resulted in the reduction of the storage modulus and the reduction of the intensity of the damping factor attributed to the segmental movements without variations in the glass transition temperature.

Rajendran Royan and collaborators (Rajendran Royan et al. 2021) published a review article exploring fused deposition additive manufacturing (FDM) technologies. The authors point to the need to overcome several challenges, such as the low mechanical properties and the printing difficulty that results when printing natural fiber composites such as jute. Throughout the article, the effects of fiber treatments, compound preparation methods, and the improvement of agents are analyzed and discussed. All these factors have already been extensively presented throughout this chapter. The mechanical properties of these composites depend on the printing parameters, so it is necessary to evaluate the effects of printing temperature, layer height, infill, and screen angle, as reported by different authors (Matsuzaki et al. 2016; Rajendran Royan et al. 2021; Valino et al. 2019).

Expectations and Remarks of Jute Fiber Composites

The applications of natural fibers composites cover a wide range of high and low-cost technologies, but composite performance depends on the intrinsic constituents' properties and manufacturing parameters.

Throughout this chapter, the applications and uses of jute fiber have been presented. Jute fiber is required for furniture to biomedical applications because of its biodegradable and nontoxic properties. The main concern for jute composites is attributed to poor bonding with polymer (thermoplastic or thermoset) matrix. Furthermore, limited fiber volume hinders it from achieving lightweight but strong and greener composites. The molecular affinity between jute fiber and polymer resins is essential to improve the overall properties of jute fiber composites. Another alternative for mechanical performance is the combination of synthetic/natural fibers. Nanoparticle reinforcement can facilitate jute-polymer bonding and increase the stiffness of composites without increasing the material density. The chemicals used in fiber treatment might cause some environmental impacts. Therefore, research in finding feasible natural biobased extracts and natural dyes for fiber treatment needs to be explored. Up today, biobased and biodegradable polymers are available. However, they are high-cost and exists some concerns about degradability.

With the traditional composite manufacturing techniques, the reliability of the quality of the jute-based composite products remains a significant issue. Disruptive composite processing techniques should be developed to obtain high-quality and consistent products with high fiber content. Improving and optimizing processing conditions should ensure composite quality without altering the cost of processing or the product cost. Some attempts point out 3D printing jute composites. However, this technology presents several observations regarding quality and time consumption. Jute sandwich panels have been found to be attractive for advanced composites in the infrastructure sector.

Natural fiber or biobased resin conveys the idea that jute fiber composites are entirely biodegradable. The preceding entails a direct step towards developing a circular economy -the circular economy approach advocates for transforming waste into resources cost-effectively-. Focusing on how to continue reusing these composites will enable decisions for sustainable product development. There are ample opportunities for research in the development of jute fiber composites. The hybridization of materials, the combination of resins, the use of related green molecules, and the continuous technological development, together with the development of new products, will allow these composites to find new markets with the offer of trendy and ecological products for the growing number of customers concerned about the environment.

References

Adak, Bapan, and Samrat Mukhopadhyay. 2016. "Jute Based All-Cellulose Composite Laminates." *Indian Journal of Fibre and Textile Research* 41(4):380–84.

Ahmed, K. Sabeel, S. Vijayarangan, and A. C. B. Naidu. 2007. "Elastic Properties, Notched Strength and Fracture Criterion in Untreated Woven Jute–Glass Fabric Reinforced Polyester Hybrid Composites." *Materials & Design* 28(8):2287–94. doi: https://doi.org/10.1016/j.matdes.2006.08.002.

Akil, Hazizan Md, Carlo Santulli, Fabrizio Sarasini, Jacopo Tirillò, and Teodoro Valente. 2014. "Environmental Effects on the Mechanical Behaviour of Pultruded Jute/Glass Fibre-Reinforced Polyester Hybrid Composites." *Composites Science and Technology* 94:62–70. doi: https://doi.org/10.1016/j.compscitech.2014.01.017.

Alim, Md Abdul, Md Moniruzzaman, Md Muzaher Hossain, Wahiduzzaman, Md Reazuddin Repon, Ismail Hossain, and Mohammad Abdul Jalil. 2022. "Manufacturing and Compatibilization of Binary Blends of Superheated Steam Treated Jute and Poly (Lactic Acid) Biocomposites by Melt-Blending Technique." *Heliyon* 8(8):e09923. doi: 10.1016/j.heliyon.2022.e09923.

Alkbir, M. F. M., S. M. Sapuan, A. A. Nuraini, and M. R. Ishak. 2016. "Fibre Properties and Crashworthiness Parameters of Natural Fibre-Reinforced Composite Structure: A Literature Review." *Composite Structures* 148:59–73. doi: 10.1016/j.compstruct.2016.01.098.

Anon. n.d. "*Fibre to Yarn.*"

Aquino, E. M. F., L. P. S. Sarmento, W. Oliveira, and R. V Silva. 2007. "Moisture Effect on Degradation of Jute/Glass Hybrid Composites." *Journal of Reinforced Plastics and Composites* 26(2):219–33. doi: 10.1177/0731684407070030.

Arao, Yoshihiko, Takayasu Fujiura, Satoshi Itani, and Tatsuya Tanaka. 2015. "Strength Improvement in Injection-Molded Jute-Fiber-Reinforced Polylactide Green-Composites." *Composites Part B: Engineering* 68:200–206. doi: https://doi.org/10.1016/j.compositesb.2014.08.032.

Ashok, R. B., C. V. Srinivasa, and B. Basavaraju. 2018. "A Review on the Mechanical Properties of Areca Fiber Reinforced Composites." *Science and Technology of Materials* 30(2):120–30. doi: https://doi.org/10.1016/j.stmat.2018.05.004.

Ashraf, Muhammad Ahsan, Mohammed Zwawi, Muhammad Taqi Mehran, Ramesh Kanthasamy, and Ali Bahadar. 2019. "Jute Based Bio and Hybrid Composites and Their Applications." *Fibers* 7(9):77. doi: 10.3390/fib7090077.

Bakhshi, Nima, and Mehdi Hojjati. 2020. "Effect of Compaction Roller on Layup Quality and Defects Formation in Automated Fiber Placement." *Journal of Reinforced Plastics and Composites* 39(1–2):3–20. doi: 10.1177/0731684419868845.

Balla, Vamsi Krishna, Kunal H. Kate, Jagannadh Satyavolu, Paramjot Singh, and Jogi Ganesh Dattatreya Tadimeti. 2019. "Additive Manufacturing of Natural Fiber Reinforced Polymer Composites: Processing and Prospects." *Composites Part B: Engineering* 174(May):106956. doi: 10.1016/j.compositesb.2019.106956.

Barkoula, N. M., Alcock, B., Cabrera, N.O., & Peijs, T. (2008). Fatigue Properties of Highly Oriented Polypropylene Tapes and All-Polypropylene Composites. *Polymers and Polymer Composites, 16*, 101 - 113. https://doi.org/10.1177/096739110801600203.

Bex, G. J., W. Six, J. De Keyzer, F. Desplentere, and A. Van Bael. 2019. "Two-Component Injection Moulding of Thermoplastics with Thermoset Rubbers: The Effect of the Mould Temperature Distribution." *AIP Conference Proceedings* 2055(January). doi: 10.1063/1.5084876.

Bi, Xiaoyu, and Runzhou Huang. 2022. "3D Printing of Natural Fiber and Composites: A State-of-the-Art Review." *Materials & Design* 222:111065. doi: https://doi.org/10.1016/j.matdes.2022.111065.

Biswas, S., Qumrul Ahsan, Ignaas Verpoest, and Mahbub Hasan. 2011. "Effect of Span Length on the Tensile Properties of Natural Fibers." *Advanced Materials Research* 264–265:445–50. doi: 10.4028/www.scientific.net/AMR.264-265.445.

Bledzki, A. K., and A. Jaszkiewicz. 2010. "Mechanical Performance of Biocomposites Based on PLA and PHBV Reinforced with Natural Fibres – A Comparative Study to PP." *Composites Science and Technology* 70(12):1687–96. doi: https://doi.org/10.1016/j.compscitech.2010.06.005.

Boopalan, M., M. Niranjanaa, and M. J. Umapathy. 2013. "Study on the Mechanical Properties and Thermal Properties of Jute and Banana Fiber Reinforced Epoxy Hybrid Composites." *Composites Part B: Engineering* 51:54–57. doi: https://doi.org/10.1016/j.compositesb.2013.02.033.

Chabros, Artur, Barbara Gawdzik, Beata Podkościelna, Martav Goliszek, and Przemyslaw Paczkowski. 2020. "Composites of Unsaturated Polyester Resins with Microcrystalline Cellulose and Its Derivatives." *Materials* 13(1). doi: 10.3390/ma13010062.

Chandekar, Harichandra, Vikas Chaudhari, and Sachin Waigaonkar. 2020. "A Review of Jute Fiber Reinforced Polymer Composites." *Materials Today: Proceedings* 26:2079–82. doi: 10.1016/j.matpr.2020.02.449.

Chauhan, Vardaan, Timo Kärki, and Juha Varis. 2019. "Review of Natural Fiber-Reinforced Engineering Plastic Composites, Their Applications in the Transportation Sector and Processing Techniques." *Journal of Thermoplastic Composite Materials*. doi: 10.1177/0892705719889095.

Chestee, Sk. Sharfuddin, Pinku Poddar, Tushar Kumar Sheel, Md. Mamunur Rashid, Ruhul A. Khan, and A. M. Sarwaruddin Chowdhury. 2017. "Short Jute Fiber Reinforced Polypropylene Composites: Effect of Nonhalogenated Fire Retardants." *Advances in Chemistry* 2017:1–8. doi: 10.1155/2017/1049513.

Chubuike, Engr. Ojukwu Martins, Chukwunyelu Christian Ebele, Engr. Ilo Fidelis Ifeanyi, Ekwueme Solomon Okwuchukwu, and Orizu Eziafa Festus. 2017. "Study on Chemical Treatments of Jute Fiber for Application in Natural Fiber Reinforced

Composites (NFRPC)." *International Journal of Advanced Engineering Research and Science* 4(2):21–26. doi: 10.22161/ijaers.4.2.4.

da Fonseca, Régis Pamponet, Janaíde Cavalcante Rocha, and Malik Cheriaf. 2021. "Mechanical Properties of Mortars Reinforced with Amazon Rainforest Natural Fibers." *Materials* 14(1). doi: 10.3390/ma14010155.

Das, Partha Pratim, and Vijay Chaudhary. 2020. "Environmental Impact and Effect of Chemical Treatment on Bio Fiber Based Polymer Composites." *Materials Today: Proceedings* 49(April):3418–22. doi: 10.1016/j.matpr.2021.03.097.

Debnath, Kishore, Inderdeep Singh, and Akshay Dvivedi. 2014. "Drilling Characteristics of Sisal Fiber-Reinforced Epoxy and Polypropylene Composites." *Materials and Manufacturing Processes* 29(11–12):1401–9. doi: 10.1080/10426914.2014.941870.

Defoirdt, Nele, Subhankar Biswas, Linde De Vriese, Le Quan Ngoc Tran, Joris Van Acker, Qumrul Ahsan, Larissa Gorbatikh, Aart Van Vuure, and Ignaas Verpoest. 2010. "Assessment of the Tensile Properties of Coir, Bamboo and Jute Fibre." *Composites Part A: Applied Science and Manufacturing* 41(5):588–95. doi: https://doi.org/10.1016/j.compositesa.2010.01.005.

Devarajan, Balaji, Rajeshkumar Lakshmi Narasimhan, Bhuvaneswari Venkateswaran, Sanjay Mavinkere Rangappa, and Suchart Siengchin. 2022. "Additive Manufacturing of Jute Fiber Reinforced Polymer Composites: A Concise Review of Material Forms and Methods." *Polymer Composites* n/a(n/a). doi: https://doi.org/10.1002/pc.26789.

Drzal, Lawrence T. n.d. *Natural Fibers, Biopolymers, and Biocompsites.*

Elanchezhian, C., B. Vijay. Ramnath, G. Ramakrishnan, M. Rajendrakumar, V. Naveenkumar, and M. K. Saravanakumar. 2018. "Review on Mechanical Properties of Natural Fiber Composites." *Materials Today: Proceedings* 5(1):1785–90. doi: 10.1016/j.matpr.2017.11.276.

Elaya Perumal, A., and N. Venkateshwaran. 2008. "Natural Fiber-Reinforced Polymer Composites in Automotive Applications- A Review." *IJAEA* 1(6):68–74.

Elkington, M., D. Bloom, C. Ward, A. Chatzimichali, and K. Potter. 2015. "Hand Layup: Understanding the Manual Process." *Advanced Manufacturing: Polymer & Composites Science* 1(3):138–51. doi: 10.1080/20550340.2015.1114801.

Ferreira, Saulo Rocha, Enzo Martinelli, Marco Pepe, Flávio De Andrade Silva, and Romildo Dias Toledo Filho. 2016. "Inverse Identification of the Bond Behavior for Jute Fibers in Cementitious Matrix." *Composites Part B: Engineering* 95:440–52. doi: 10.1016/j.compositesb.2016.03.097.

Fiore, Vincenzo, and Luigi Calabrese. 2019. "Effect of Stacking Sequence and Sodium Bicarbonate Treatment on Quasi-Static and Dynamic Mechanical Properties of Flax/Jute Epoxy-Based Composites." *Materials* 12(9). doi: 10.3390/ma12091363.

Franco-Urquiza, E. A. 2022. "Applications and Drawbacks of Epoxy/Synthetic/Natural Fiber Hybrid Composites." Pp. 1121–54 in *Handbook of Epoxy/Fiber Composites*, edited by S. Mavinkere Rangappa, J. Parameswaranpillai, S. Siengchin, and S. Thomas. Singapore: Springer Nature Singapore.

Franco-Urquiza, E. A., V. Renteria, R. Perez-Mora, P. González-García, and M. Torres-Arellano. 2020. "Mode i Interlaminar Fracture Toughness of Biolaminates Composites Charged with Reinforced Particles." in *Composites and Advanced Materials Expo, CAMX 2020.*

Franco-Urquiza, E. A., R. S. Saleme-Osornio, and R. Ramírez-Aguilar. 2021. "Mechanical Properties of Hybrid Carbonized Plant Fibers Reinforced Bio-Based Epoxy Laminates." *Polymers* 13(19). doi: 10.3390/polym13193435.

Franco-Urquiza, E. A., B. Vázquez, and C. A. Escalante-Velázquez. 2021. "Análisis Del Efecto de Capilaridad En El Procesos de Infusión de Bio-Resinas En Tejidos de Fibra Natural [Analysis of the Effect of Capillarity in the Infusion Processes of Bio-Resins in Natural Fiber Fabrics]." *Afinidad* 78(594):238–44.

Franco-Urquiza, E., J. G. Perez, M. Sánchez-Soto, O. O. Santana, and M. L. Maspoch. 2010. "The Effect of Organo-Modifier on the Structure and Properties of Poly[Ethylene-(Vinyl Alcohol)]/Organo-Modified Montmorillonite Composites." *Polymer International* 59(6):778–86. doi: 10.1002/pi.2788.

Franco-Urquiza, E. A., and Rentería-Rodríguez, A. V. 2021. "Effect of Nanoparticles on the Mechanical Properties of Kenaf Fiber-Reinforced Bio-Based Epoxy Resin." *Textile Research Journal* 91(11–12):1313–25. doi: 10.1177/0040517520980459.

Franco-Urquiza, Edgar Adrián. 2021. "Clay-Based Polymer Nanocomposites: Essential Work of Fracture." *Polymers* 1–42.

Franco-Urquiza, Edgar Adrian, Jonathan Cailloux, Orlando Santana, Maria Lluisa Maspoch, and Julio Cesar Velazquez Infante. 2015. "The Influence of the Clay Particles on the Mechanical Properties and Fracture Behavior of PLA/o-MMT Composite Films." *Advances in Polymer Technology* 34(1):n/a-n/a. doi: 10.1002/adv.21470.

Franco-Urquiza, Edgar Adrián, Yael Ramírez Escamilla, and Perla Itzel Alcántara Llanas. 2021. "Characterization of 3D Printing on Jute Fabrics." *Polymers* 13(19):3202. doi: 10.3390/polym13193202.

Franco-Urquiza, Edgar Adrián, Raúl Samir Saleme-Osornio, and Rodrigo Ramírez-Aguilar. 2021. "Mechanical Properties of Hybrid Carbonized Plant Fibers Reinforced Bio-Based Epoxy Laminates." *Polymers* 13(19):3435. doi: 10.3390/polym13193435.

Ghani, Muhammad Usman, Amna Siddique, Kahsay Gebresilassie Abraha, Lan Yao, Wei Li, Muhammad Qamar Khan, and Ick-Soo Kim. 2022. "Performance Evaluation of Jute/Glass-Fiber-Reinforced Polybutylene Succinate (PBS) Hybrid Composites with Different Layering Configurations." *Materials* 15(3). doi: 10.3390/ma15031055.

Gon, Debiprasad, Kousik Das, Palash Paul, and Subhankar Maity. 2012. "Jute Composites as Wood Substitute." *International Journal of Textile Science* 1:84–93. doi: 10.5923/j.textile.20120106.05.

Grande, C., and F. G. Torres. 2005. "Investigation of Fiber Organization and Damage during Single Screw Extrusion of Natural Fiber Reinforced Thermoplastics." *Advances in Polymer Technology* 24(2):145–56. doi: https://doi.org/10.1002/adv.20037.

Guha Roy, T. K., A. K. Mukhopadhyay, and A. K. Mukherjee. 1984. "Surface Features of Jute Fiber Using Scanning Electron Microscopy." *Textile Research Journal* 54(12):874–82. doi: 10.1177/004051758405401214.

Gunning, Michael A., Luke M. Geever, John A. Killion, John G. Lyons, and Clement L. Higginbotham. 2013. "The Effect of Processing Conditions for Polylactic Acid Based Fibre Composites via Twin-Screw Extrusion." *Journal of Reinforced Plastics and Composites* 33(7):648–62. doi: 10.1177/0731684413512225.

Guo, Aofei, Zhihui Sun, and Jagannadh Satyavolu. 2019. "Impact of Chemical Treatment on the Physiochemical and Mechanical Properties of Kenaf Fibers." *Industrial Crops and Products* 141:111726. doi: https://doi.org/10.1016/j.indcrop.2019.111726.

Hasan, Mahmudul, Abu Saifullah, Hom N. Dhakal, Shahjalal Khandaker, and Forkan Sarker. 2021. "Improved Mechanical Performances of Unidirectional Jute Fibre Composites Developed with New Fibre Architectures." *RSC Advances* 11(37):23010–22. doi: 10.1039/D1RA03515K.

Hashemi, S. 2002. "Influence of Temperature on Weldline Strength of Injection Moulded Short Glass Fibre Styrene Maleic Anhydride Polymer Composites." *Plastics, Rubber and Composites* 31(7):318–24. doi: 10.1179/146580102225005027.

Hossen, Md. Faruk, Md. Ali Asraf, Md. Kudrat-E. Zahan, Md. Masuqul Haque, Rausan Zamir, and Choudhury M. Zakaria. 2020. "Investigation of the Physico-Chemical Absorption Characterizations of Jute Polymer Clay Nanocomposites as a Function of Chemical Treatments." *International Research Journal of Pure and Applied Chemistry* 21(10):118–30. doi: 10.9734/IRJPAC/2020/v21i1030214.

Huda, M. S., L. T. Drzal, D. Ray, A. K. Mohanty, and M. Mishra. 2008. "Natural-Fiber Composites in the Automotive Sector." Pp. 221–68 in *Properties and Performance of Natural-Fibre Composites*.

Huq, Tanzina, Avik Khan, Farah M. J. Hossain, Tahmina Akter, Haydar U. Zaman, Nousin Aktar, Mohammad O. Tuhin, Tuhidul Islam, and Ruhul A. Khan. 2013. "Gamma-Irradiated Jute/Polypropylene Composites by Extrusion Molding." *Composite Interfaces* 20(2):93–105. doi: 10.1080/15685543.2013.762741.

Jagadish, and Bhowmik Sumit. 2021. *Manufacturing and Processing of Natural Filler Based Polymer Composites*.

Jawaid, M., H. P. S. Abdul Khalil, and A. Abu Bakar. 2010. "Mechanical Performance of Oil Palm Empty Fruit Bunches/Jute Fibres Reinforced Epoxy Hybrid Composites." *Materials Science and Engineering: A* 527(29):7944–49. doi: https://doi.org/10.1016/j.msea.2010.09.005.

Jawaid, M., H. P. S. Abdul Khalil, A. Abu Bakar, and P. Noorunnisa Khanam. 2011. "Chemical Resistance, Void Content and Tensile Properties of Oil Palm/Jute Fibre Reinforced Polymer Hybrid Composites." *Materials & Design* 32(2):1014–19. doi: https://doi.org/10.1016/j.matdes.2010.07.033.

Jeong, Euichul, Yongdae Kim, Seokkwan Hong, Kyunghwan Yoon, and Sunghee Lee. 2022. "Innovative Injection Molding Process for the Fabrication of Woven Fabric Reinforced Thermoplastic Composites." *Polymers* 14(8). doi: 10.3390/polym14081577.

Jha, Kanishka, Bibhuti Bhusan Samantaray, and Paresh Tamrakar. 2018. "A Study on Erosion and Mechanical Behavior of Jute/E-Glass Hybrid Composite." *Materials Today: Proceedings* 5(2, Part 1):5601–7. doi: https://doi.org/10.1016/j.matpr.2017.12.151.

Katayama, T., K. Tanaka, T. Murakami, and K. Uno. 2006. "Compression Moulding of Jute Fabric Reinforced Thermoplastic Composites Based on PLA Non-Woven Fabric." *WIT Transactions on the Built Environment* 85:159–67. doi: 10.2495/HPSM06017.

Khan, GM Arifuzzaman, and Md Shamsul Alam. 2016. "Surface Chemical Treatments of Jute Fiber for High Value Composite Uses." *Research & Reviews: Journal of Material Sciences* 01(02). doi: 10.4172/2321-6212.1000110.

Khan, Ruhul A., Mubarak A. Khan, Haydar U. Zaman, Fahmida Parvin, Towhidul Islam, Farah Nigar, Rafiqul Islam, Suvasree Saha, and A. I. Mustafa. 2012. "Fabrication and Characterization of Jute Fabric-Reinforced PVC-Based Composite." *Journal of Thermoplastic Composite Materials* 25(1):45–58. doi: 10.1177/0892705711404726.

Khan, T., M. T. H. Sultan, A. U. M. Shah, A. H. Ariffin, and M. Jawaid. 2021. "The Effects of Stacking Sequence on the Tensile and Flexural Properties of Kenaf/Jute Fibre Hybrid Composites." *Journal of Natural Fibers* 18(3):452–63. doi: 10.1080/15440478.2019.1629148.

Kumar, Asheesh, and Anshuman Srivastava. 2017. "Preparation and Mechanical Properties of Jute Fiber Reinforced Epoxy Composites." *Industrial Engineering & Management* 06(04):4–7. doi: 10.4172/2169-0316.1000234.

Kundu, B. C. (Balai Chand). 1959. *Jute in India* / by B. C. Kundu, K. C. Basak [and] P. B. Sarkar. edited by K. C. (Kiran C. Basak, P. B. Sarkar, and I. C. J. Committee. Calcutta: Indian Central Jute Committee.

Laraba, Selsabil Rokia, Amine Rezzoug, Rafik Halimi, Luo Wei, Yuhao yang, Said Abdi, Yulin Li, and Wei Jie. 2022. "Development of Sandwich Using Low-Cost Natural Fibers: Alfa-Epoxy Composite Core and Jute/Metallic Mesh-Epoxy Hybrid Skin Composite." *Industrial Crops and Products* 184:115093. doi: https://doi.org/10.1016/j.indcrop.2022.115093.

Le Duigou, Antoine, David Correa, Masahito Ueda, Ryosuke Matsuzaki, and Mickael Castro. 2020. "A Review of 3D and 4D Printing of Natural Fibre Biocomposites." *Materials & Design* 194:108911. doi: https://doi.org/10.1016/j.matdes.2020.108911.

Mahesh, Vinyas, Vishwas Mahesh, and Dineshkumar Harursampath. 2021. "Physio-Mechanical Characterization of Jute/Kevlar Hybrid Composites Coupled with MADM Approach for Selection of Composites." *Journal of Natural Fibers* 1–9. doi: 10.1080/15440478.2021.2009403.

Masoodi, Reza, and Krishna M. Pillai. 2012. "A Study on Moisture Absorption and Swelling in Bio-Based Jute-Epoxy Composites." *Journal of Reinforced Plastics and Composites* 31(5):285–94. doi: 10.1177/0731684411434654.

Maspoch, Maria L., Edgar Franco-Urquiza, José Gamez-Perez, Orlando O. Santana, and Miguel Sánchez-Soto. 2009. "Fracture Behaviour of Poly[Ethylene-(Vinyl Alcohol)]/Organo-Clay Composites." *Polymer International* 58(6):648–55. doi: 10.1002/pi.2574.

Mathur, Manisha. 2021. "Natural Fiber Composites." *Man-Made Textiles in India* 49(2):1. doi: 10.1201/9781351050944-1.

Matsuzaki, Ryosuke, Masahito Ueda, Masaki Namiki, Tae-Kun Jeong, Hirosuke Asahara, Keisuke Horiguchi, Taishi Nakamura, Akira Todoroki, and Yoshiyasu Hirano. 2016. "Three-Dimensional Printing of Continuous-Fiber Composites by in-Nozzle Impregnation." *Scientific Reports* 6(1):1–7.

Memon, Anin, and Asami Nakai. 2013. "Fabrication and Mechanical Properties of Jute Spun Yarn/PLA Unidirection Composite by Compression Molding." *Energy Procedia* 34:830–38. doi: 10.1016/j.egypro.2013.06.819.

Miah, M. J., M. A. Khan, and R. A. Khan. 2011. "Fabrication and Characterization of Jute Fiber Reinforced Low Density Polyethylene Based Composites: Effects of Chemical Treatment." *Journal of Scientific Research* 3(2):249–59. doi: 10.3329/jsr.v3i2.6763.

Miyagawa, Hiroaki, Robert J. Jurek, Amar K. Mohanty, Manjusri Misra, and Lawrence T. Drzal. 2006. "Biobased Epoxy/Clay Nanocomposites as a New Matrix for CFRP." *Composites Part A: Applied Science and Manufacturing* 37(1):54–62. doi: 10.1016/j.compositesa.2005.05.014.

Mohan, Rengarajan, and Kishore. 1985. "Jute-Glass Sandwich Composites." *Journal of Reinforced Plastics and Composites* 4(2):186–94. doi: 10.1177/073168448500400203.

Murdani, A., and U. S. Amrullah. 2021. "Flexural Behaviour of Jute, Glass, and Carbon Fibre Reinforced Polyester Hybrid Composites." *IOP Conference Series: Materials Science and Engineering* 1173(1):012067. doi: 10.1088/1757-899x/1173/1/012067.

Muzammil, Muhammad, Mohsin Ejaz, Rehman Shah, S. Kamran Afaq, and Jung-il Song. 2022. "A Bio-Based Approach to Simultaneously Improve Flame Retardancy, Thermal Stability and Mechanical Properties of Nano-Silica Filled Jute / Thermoplastic Starch Composite." *Materials Chemistry and Physics* 289(July): 126485. doi: 10.1016/j.matchemphys.2022.126485.

Nayak, Subhakanta, and Jyoti Ranjan Mohanty. 2019. "Influence of Chemical Treatment on Tensile Strength, Water Absorption, Surface Morphology, and Thermal Analysis of Areca Sheath Fibers." *Journal of Natural Fibers* 16(4):589–99.

Niloy Rahaman, Md., Md. Sahadat Hossain, Md. Razzak, Muhammad B. Uddin, A. M. Sarwaruddi. Chowdhury, and Ruhul A. Khan. 2019. "Effect of Dye and Temperature on the Physico-Mechanical Properties of Jute/PP and Jute/LLDPE Based Composites." *Heliyon* 5(6):e01753. doi: https://doi.org/10.1016/j.heliyon.2019.e01753.

Nurul Hidayah, I., D. Nuur Syuhada, H. P. S. Abdul Khalil, Z. A. M. Ishak, and M. Mariatti. 2019. "Enhanced Performance of Lightweight Kenaf-Based Hierarchical Composite Laminates with Embedded Carbon Nanotubes." *Materials & Design* 171:107710. doi: 10.1016/j.matdes.2019.107710.

Ogorodnyk, Olga, and Kristian Martinsen. 2018. "Monitoring and Control for Thermoplastics Injection Molding A Review." *Procedia CIRP* 67:380–85. doi: https://doi.org/10.1016/j.procir.2017.12.229.

Pailoor, Sandhyarani, H. N. Narasimha Murthy, and T. N. Sreenivasa. 2021. "Drilling of In-Line Compression Molded Jute / Polypropylene Composites." *Journal of Natural Fibers* 18(1):91–104. doi: 10.1080/15440478.2019.1612309.

Pandita, Surya D., Xiaowen Yuan, Munirah A. Manan, Chun H. Lau, Alamelu S. Subramanian, and Jun Wei. 2013. "Evaluation of Jute/Glass Hybrid Composite Sandwich: Water Resistance, Impact Properties and Life Cycle Assessment." *Journal of Reinforced Plastics and Composites* 33(1):14–25. doi: 10.1177/0731684413505349.

Parbin, Shahana, Nitin Kumar Waghmare, Suraj Kumar Singh, and Sabah Khan. 2019. "Mechanical Properties of Natural Fiber Reinforced Epoxy Composites: A Review." *Procedia Computer Science* 152:375–79. doi: 10.1016/j.procs.2019.05.003.

Paul, Tamal Krishna, Tazin Ibna Jalil, Md. Shohan Parvez, Md. Reazuddin Repon, Ismail Hossain, Md. Abdul Alim, Tarikul Islam, and Mohammad Abdul Jalil. 2022. "A Prognostic Based Fuzzy Logic Method to Speculate Yarn Quality Ratio in Jute Spinning Industry." *Textiles* 2(3):422–35. doi: 10.3390/textiles2030023.

Pawar, Manisha, Aparna Kadam, Omprakash Yemul, Viresh Thamke, and Kisan Kodam. 2016. "Biodegradable Bioepoxy Resins Based on Epoxidized Natural Oil (Cottonseed & Algae) Cured with Citric and Tartaric Acids through Solution Polymerization: A Renewable Approach." *Industrial Crops and Products* 89:434–47. doi: 10.1016/j.indcrop.2016.05.025.

Puglia, D., J. Biagiotti, and J. M. Kenny. 2005. "A Review on Natural Fibre-Based Composites—Part II." *Journal of Natural Fibers* 1(3):23–65. doi: 10.1300/ J395v 01n03_03.

Rabbi, M. S., Tansirul Islam, and G. M. Sadiqu. Islam. 2021. "Injection-Molded Natural Fiber-Reinforced Polymer Composites–a Review." *International Journal of Mechanical and Materials Engineering* 16(1):1–21. doi: 10.1186/s40712-021-00139-1.

Rajasekhar, P., G. Ganesan, and C. Senthilkumar. 2014. "Studies on Tribological Behavior of Polyamide Filled Jute Fiber-Nano-Zno Hybrid Composites." *Procedia Engineering* 97(December):2099–2109. doi: 10.1016/j.proeng.2014.12.453.

Rajendran Royan, Nishata Royan, Jie Sheng Leong, Wai Nam Chan, Jie Ren Tan, and Zainon Sharmila Binti Shamsuddin. 2021. "Current State and Challenges of Natural Fibre-Reinforced Polymer Composites as Feeder in Fdm-Based 3d Printing." *Polymers* 13(14). doi: 10.3390/polym13142289.

Rajesh, Gunti, and Atluri V. Ratna Prasad. 2014. "Tensile Properties of Successive Alkali Treated Short Jute Fiber Reinforced PLA Composites." *Procedia Materials Science* 5:2188–96. doi: 10.1016/j.mspro.2014.07.425.

Rajole, Sangamesh, K. S. Ravishankar, and S. M. Kulkarni. 2020. "Performance Study of Jute-Epoxy Composites/Sandwiches under Normal Ballistic Impact." *Defence Technology* 16(4):947–55. doi: 10.1016/j.dt.2019.11.011.

Ramamoorthy, Sunil Kumar, Mikael Skrifvars, and Anders Persson. 2015. "A Review of Natural Fibers Used in Biocomposites: Plant, Animal and Regenerated Cellulose Fibers." *Polymer Reviews* 55(1):107–62. doi: 10.1080/15583724.2014.971124.

Ramprasath, B., R. Murugesan, Abhik Banerjee, Abhinav Anand, and Shashank. 2020. "A Comparative Study of Sandwich and Hybrid Sandwich Composites Using Jute and Kevlar Fibers." *IOP Conference Series: Materials Science and Engineering* 912(5). doi: 10.1088/1757-899X/912/5/052031.

Rana, A. K., and K. Jayachandran. 2000. "Jute Fiber for Reinforced Composites and Its Prospects." *Molecular Crystals and Liquid Crystals Science and Technology. Section A. Molecular Crystals and Liquid Crystals* 353(1):35–45. doi: 10.1080/1058725 0008025646.

Ranganathan, Nalini, Kristiina Oksman, Sanjay K. Nayak, and Mohini Sain. 2015. "Regenerated Cellulose Fibers as Impact Modifier in Long Jute Fiber Reinforced Polypropylene Composites: Effect on Mechanical Properties, Morphology, and Fiber Breakage." *Journal of Applied Polymer Science* 132(3). doi: https://doi.org/10.1002/app.41301.

Rangappa, Sanjay Mavinkere, Jyotishkumar Parameswaranpillai, Suchart Siengchin, and Sabu Thomas. n.d. *Handbook of Epoxy / Fiber Composites*.

Ray, Dipa, B. K. Sarkar, R. K. Basak, and A. K. Rana. 2004. "Thermal Behavior of Vinyl Ester Resin Matrix Composites Reinforced with Alkali-Treated Jute Fibers." *Journal of Applied Polymer Science* 94(1):123–29. doi: 10.1002/app.20754.

Robson, S., and T. C. Goodhead. 2015. "A Process for Incorporating Automotive Shredder Residue into Thermoplastic Mouldings." 139(2003):327–31. doi: 10.1016/S0924-0136(03)00549-1.

Rodriguez, Exequiel S., Pablo M. Stefani, and Analia Vazquez. 2007. "Effects of Fibers' Alkali Treatment on the Resin Transfer Molding Processing and Mechanical Properties of Jute—Vinylester Composites." *Journal of Composite Materials* 41(14):1729–41. doi: 10.1177/0021998306069889.

Roy, A. K., S. C. Bag, D. Sardar, and S. K. Sen. 1991. "Infrared Spectra of Jute Stick Bleached with Sodium Chlorite and Hydrogen Peroxide." *Journal of Applied Polymer Science* 43(12):2187–92. doi: 10.1002/app.1991.070431205.

Roy, A. K., S. K. Sen, S. C. Bag, and S. N. Pandey. 1991. "Scanning Electron Microscopic Study of Jute Stick Treated with Some Pulping and Bleaching Agents." *Holzforschung* 45(3):209–14. doi: 10.1515/hfsg.1991.45.3.209.

Roy, Juganta K., Ruhul A. Khan, K. M. Zakir Hossain, Mubarak A. Khan, Sabyasachy Mistry, and A. M. Sarwaruddin Chowdhury. 2012. "Effect of Sodium Bicarbonate on the Mechanical and Degradation Properties of Short Jute Fiber Reinforced Polypropylene Composite by Extrusion Technique." *International Journal of Polymeric Materials and Polymeric Biomaterials* 61(8):571–86. doi: 10.1080/00914037.2011.610039.

Rubin, Irvin I. 1991. *"Injection Molding of Thermoplastics BT - SPI Plastics Engineering Handbook of the Society of the Plastics Industry, Inc."* Pp. 133–78 in, edited by M. L. Berins. Boston, MA: Springer US.

Saaidia, Aziz, Abderrezak Bezazi, Ahmed Belbah, Salah Amroune, and Fabrizio Scarpa. 2015. "Evaluation of Mechanical Properties of Jute Yarns by Two- and Three-Parameters Weibull Method." *Structural Integrity and Life* 15(3):157–62.

Safiee, Sahnizam, Hazizan M. D. Akil, Adlan Akram Mohammad Mazuki, Zainal Ariffin Mohd Ishak, and Azhar Abu Bakar. 2011. "Properties of Pultruded Jute Fiber Reinforced Unsaturated Polyester Composites." *Advanced Composite Materials* 20(3):231–44. doi: 10.1163/092430410X547047.

Salman, Suhad D. 2020. "Effects of Jute Fibre Content on the Mechanical and Dynamic Mechanical Properties of the Composites in Structural Applications." *Defence Technology* 16(6):1098–1105. doi: 10.1016/j.dt.2019.11.013.

Samanta, Ashis Kumar, Asis Mukhopadhyay, and Swapan Kumar Ghosh. 2020. *Processing of Jute Fibres and Its Applications*.

Sanvezzo, Paula Bertolino, and Marcia Cristina Branciforti. 2021. "Recycling of Industrial Waste Based on Jute Fiber-Polypropylene: Manufacture of Sustainable Fiber-Reinforced Polymer Composites and Their Characterization before and after Accelerated Aging." *Industrial Crops and Products* 168(November 2020):113568. doi: 10.1016/j.indcrop.2021.113568.

Sarasook, Prattakon, Putinun Uawongsuwan, Anin Memon, and Hiroyuki Hamada. 2020. "Jute Fiber Reinforced Thermoplastic Composites Fabricated by Direct Fiber Feeding Injection Molding (DFfim) Process." *Key Engineering Materials* 856 KEM:268–75. doi: 10.4028/www.scientific.net/KEM.856.268.

Sathishkumar, S., A. V. Suresh, M. Nagamadhu, and M. Krishna. 2017. "The Effect of Alkaline Treatment on Their Properties of Jute Fiber Mat and Its Vinyl Ester Composites." *Materials Today: Proceedings* 4(2):3371–79. doi: 10.1016/j.matpr.2017.02.225.

Sathishkumar, T. P., L. Rajeshkumar, G. Rajeshkumar, M. R. Sanjay, Suchart Siengchin, and Navanee Thakrishnan. 2022. "Improving the Mechanical Properties of Jute Fiber Woven Mat Reinforced Epoxy Composites with Addition of Zinc Oxide Filler." *E3S Web of Conferences* 355:02006. doi: 10.1051/e3sconf/202235502006.

Sathiyamoorthy, Margabandu, and Subramaniam Senthilkumar. 2020. "Mechanical, Thermal, and Water Absorption Behaviour of Jute/Carbon Reinforced Hybrid Composites." *Sadhana - Academy Proceedings in Engineering Sciences* 45(1):1–12. doi: 10.1007/s12046-020-01514-y.

Saw, Sudhir Kumar, and Chandan Datta. 2009. "Thermomechanical Properties of Jute/Bagasse Hybrid Fibre Reinforced Epoxy Thermoset Composites." *BioResources* 4(4):1455–76.

Selvan, S. Panneer. 2021. "A Comparison Study on Delamination in Hand Layup and Compression Moulded Jute - *Human Hair Polymer Composites*." 8(4):10585–92.

Selver, Erdem, Nuray Ucar, and Turgut Gulmez. 2018. "Effect of Stacking Sequence on Tensile, Flexural and Thermomechanical Properties of Hybrid Flax/Glass and Jute/Glass Thermoset Composites." *Journal of Industrial Textiles* 48(2):494–520. doi: 10.1177/1528083717736102.

Shahinur, Sweety, M. M. Alamgir Sayeed, Mahbub Hasan, Abu Sadat Muhammad Sayem, Julfikar Haider, and Sharifu Ura. 2022. "Current Development and Future Perspective on Natural Jute Fibers and Their Biocomposites." *Polymers* 14(7). doi: 10.3390/polym14071445.

Shahinur, Sweety, Mahbub Hasan, Qumrul Ahsan, and Julfikar Haider. 2020. "Effect of Chemical Treatment on Thermal Properties of Jute Fiber Used in Polymer Composites." *Journal of Composites Science* 4(3). doi: 10.3390/jcs4030132.

Shahinur, Sweety, Mahbub Hasan, Qumrul Ahsan, Dilip Kumar Saha, and Saiful Islam. 2015. "*Characterization on the Properties of Jute Fiber at Different Portions*." 2015.

Shanmugam, D., and M. Thiruchitrambalam. 2013. "Static and Dynamic Mechanical Properties of Alkali Treated Unidirectional Continuous Palmyra Palm Leaf Stalk Fiber/Jute Fiber Reinforced Hybrid Polyester Composites." *Materials & Design* 50:533–42. doi: https://doi.org/10.1016/j.matdes.2013.03.048.

Shivamurthy, B., Nithesh Naik, B. H. S. Thimappa, and Ritesh Bhat. 2020. "Mechanical Property Evaluation of Alkali-Treated Jute Fiber Reinforced Bio-Epoxy Composite Materials." *Materials Today: Proceedings* 28:2116–20. doi: 10.1016/j.matpr.2020.04.016.

Singh, Manoj Kumar, and Sunny Zafar. 2020. "Effect of Layering Sequence on Mechanical Properties of Woven Kenaf/Jute Fabric Hybrid Laminated Microwave-Processed Composites." *Journal of Industrial Textiles*. doi: 10.1177/1528083720911219.

Singh, Manoj Kumar, and Sunny Zafar. 2022. "Effect of Layering Sequence on Mechanical Properties of Woven Kenaf/Jute Fabric Hybrid Laminated Microwave-Processed Composites." *Journal of Industrial Textiles* 51(2):2731S-2752S. doi: 10.1177/1528083720911219.

Singh, Vivek, Parshant Kumar, and V. K. Srivastava. 2022. "Effect of Particle Doping on the Mechanical Behavior of 2D Woven (0 ∘ /90 ∘) Jute Fabric (Plain Weave) Reinforced Polymer Matrix Composites." *Sadhana - Academy Proceedings in Engineering Sciences* 47(3). doi: 10.1007/s12046-022-01883-6.

Suddell, Brett, and William Evans. 2010. "Natural Fiber Composites in Automotive Applications." in *Natural Fibers, Biopolymers, and Biocomposites*.

Sun, Zhan-Ying, Hai-Shan Han, and Gan-Ce Dai. 2009. "Mechanical Properties of Injection-Molded Natural Fiber-Reinforced Polypropylene Composites: Formulation and Compounding Processes." *Journal of Reinforced Plastics and Composites* 29(5):637–50. doi: 10.1177/0731684408100264.

Taylor, Publisher, Michael Karus, and Markus Kaup. 2002. "Natural Fiber Composites in the European Automotive Industry." *Journal of Industrial Hemp* 126(January 2015):9-12 ST-Natural fiber composites in the Europea. doi: 10.1300/J237v07n01.

Thiagamani, Senthil Muthu Kumar, Senthilkumar Krishnasamy, Chandrasekar Muthukumar, Jiratti Tengsuthiwat, Rajini Nagarajan, Suchart Siengchin, and Sikiru O. Ismail. 2019. "Investigation into Mechanical, Absorption and Swelling Behaviour of Hemp/Sisal Fibre Reinforced Bioepoxy Hybrid Composites: Effects of Stacking Sequences." *International Journal of Biological Macromolecules* 140:637–46. doi: 10.1016/j.ijbiomac.2019.08.166.

Thyavihalli Girijappa, Yashas Gowda, Sanjay Mavinkere Rangappa, Jyotishkumar Parameswaranpillai, and Suchart Siengchin. 2019. "Natural Fibers as Sustainable and Renewable Resource for Development of Eco-Friendly Composites: A Comprehensive Review." *Frontiers in Materials* 6:226. doi: 10.3389/fmats.2019.00226.

Thygesen, Anders, Bo Madsen, Anne Belinda Bjerre, and Hans Lilholt. 2011. "Cellulosic Fibers: Effect of Processing on Fiber Bundle Strength." *Journal of Natural Fibers* 8(3):161–75. doi: 10.1080/15440478.2011.602236.

Torrado, Angel R., Corey M. Shemelya, Joel D. English, Yirong Lin, Ryan B. Wicker, and David A. Roberson. 2015. "Characterizing the Effect of Additives to ABS on the Mechanical Property Anisotropy of Specimens Fabricated by Material Extrusion 3D Printing." *Additive Manufacturing* 6:16–29. doi: https://doi.org/10.1016/j.addma.2015.02.001.

Torres-Arellano, M., V. Renteria-Rodríguez, and E. Franco-Urquiza. 2020. "Mechanical Properties of Natural-Fiber-Reinforced Biobased Epoxy Resins Manufactured by Resin Infusion Process." *Polymers* 12(12):1–17. doi: 10.3390/polym12122841.

Torres-Arellano, Mauricio, Victoria Renteria-Rodríguez, and Edgar Franco-Urquiza. 2020. "Mechanical Properties of Natural-Fiber-Reinforced Biobased Epoxy Resins Manufactured by Resin Infusion Process." *Polymers* 12(12):2841. doi: 10.3390/polym12122841.

Torres, Mauricio, Victoria Renteria Rodriguez, Perla Itzel Alcantara, and Edgar Franco-Urquiza. 2020. "Mechanical Properties and Fracture Behaviour of Agave Fibers Bio-

Based Epoxy Laminates Reinforced with Zinc Oxide." *Journal of Industrial Textiles* 152808372096568. doi: 10.1177/1528083720965689.

Valino, Arnaldo D., John Ryan C. Dizon, Alejandro H. Espera, Qiyi Chen, Jamie Messman, and Rigoberto C. Advincula. 2019. "Advances in 3D Printing of Thermoplastic Polymer Composites and Nanocomposites." *Progress in Polymer Science* 98:101162. doi: https://doi.org/10.1016/j.progpolymsci.2019.101162.

van den Oever, M. J. A., and M. H. B. Snijder. 2008. "Jute Fiber Reinforced Polypropylene Produced by Continuous Extrusion Compounding, Part 1: Processing and Ageing Properties." *Journal of Applied Polymer Science* 110(2):1009–18. doi: https://doi.org/10.1002/app.28682.

van Oosterom, S., T. Allen, M. Battley, and S. Bickerton. 2019. "An Objective Comparison of Common Vacuum Assisted Resin Infusion Processes." *Composites Part A: Applied Science and Manufacturing* 125:105528. doi: https://doi.org/10. 1016/j.compositesa.2019.105528.

Velasco-Parra, J. A., B. A. Ramón-Valencia, and W. J. Mora-Espinosa. 2021. "Mechanical Characterization of Jute Fiber-Based Biocomposite to Manufacture Automotive Components." *Journal of Applied Research and Technology* 19(5):472–91. doi: 10.22201/icat.24486736e.2021.19.5.1220.

Verhagen, M. A., T. Duelge, K. M. Pillai, and R. Masoodi. 2009. "A Study of Change in Properties of Polymer Composite after the Replacement of Glass with Jute Fibers." *24th Annual Technical Conference of the American Society for Composites 2009 and 1st Joint Canadian-American Technical Conference on Composites* 4 (December 2014):2738–45.

Viju, Subramoniapillai, and G. Thilagavathi. 2022. "Characterization of Surface Modified Nettle Fibers for Composite Reinforcement." *Journal of Natural Fibers* 19(5):1819–27. doi: 10.1080/15440478.2020.1788491.

Wang, Hua, Hafeezullah Memon, Elwathig A. M. Hassan, Sohag Miah, and Arshad Ali. n.d. "Effect of Jute Fiber Modification on Mechanical Properties of Jute Fiber Composite." *Materials* (Basel). 12(8): 1226.

Westman, Matthew P., Leonard S. Fifield, Kevin L. Simmons, Sachin Laddha, and Tyler A. Kafentzis. 2010. *Natural Fiber Composites: A Review.* U.S. Dept. of Energy.

Xian, Guijun, Rui Guo, and Chenggao Li. 2022. "Combined Effects of Sustained Bending Loading, Water Immersion and Fiber Hybrid Mode on the Mechanical Properties of Carbon/Glass Fiber Reinforced Polymer Composite." *Composite Structures* 281:115060. doi: https://doi.org/10.1016/j.compstruct.2021.115060.

Yang, Yuqiu, Tomoko Ota, Tohru Morii, and Hiroyuki Hamada. 2011. "Mechanical Property and Hydrothermal Aging of Injection Molded Jute/Polypropylene Composites." *Journal of Materials Science* 46(8):2678–84. doi: 10.1007/s10853-010-5134-8.

Yorseng, Krittirash, Sanjay Mavinkere Rangappa, Harikrishnan Pulikkalparambil, Suchart Siengchin, and Jyotishkumar Parameswaranpillai. 2020. "Accelerated Weathering Studies of Kenaf/Sisal Fiber Fabric Reinforced Fully Biobased Hybrid Bioepoxy Composites for Semi-Structural Applications: Morphology, Thermo-Mechanical, Water Absorption Behavior and Surface Hydrophobicity." *Construction and Building Materials* 235:117464. doi: 10.1016/j.conbuildmat.2019.117464.

Zafar, Mohammad Tahir, Sanjeev Kumar, Rajendra Kumar Singla, Saurindra Nath Maiti, and Anup Kumar Ghosh. 2018. "Surface Treated Jute Fiber Induced Foam Microstructure Development in Poly(Lactic Acid)/Jute Fiber Biocomposites and Their Biodegradation Behavior." *Fibers and Polymers* 19(3):648–59. doi: 10.1007/s12221-018-7428-4.

Zafar, Mohammad Tahir, Saurindra Nath Maiti, and Anup Kumar Ghosh. 2016. "Effect of Surface Treatments of Jute Fibers on the Microstructural and Mechanical Responses of Poly(Lactic Acid)/Jute Fiber Biocomposites." *RSC Advances* 6(77):73373–82. doi: 10.1039/C6RA17894D.

Zamri, Mohd Hafiz, Hazizan Md Akil, Azhar Abu Bakar, Zainal Arifin Mohd Ishak, and Leong Wei Cheng. 2011. "Effect of Water Absorption on Pultruded Jute/Glass Fiber-Reinforced Unsaturated Polyester Hybrid Composites." *Journal of Composite Materials* 46(1):51–61. doi: 10.1177/0021998311410488.

Zhang, Yuping, Xungai Wang, Ning Pan, and R. Postle. 2002. "Weibull Analysis of the Tensile Behavior of Fibers with Geometrical Irregularities." *Journal of Materials Science* 37(7):1401–6. doi: 10.1023/A:1014580814803.

Chapter 2

Chemical Modifications of Jute Fiber Properties for Lifecycle Enhancement by Utilizing in Wastewater Treatment

Aleksandra Ivanovska[1,*] and Mirjana Kostic[2]

[1]University of Belgrade, Innovation Center of the Faculty of Technology and Metallurgy, Belgrade, Serbia
[2]University of Belgrade, Faculty of Technology and Metallurgy, Belgrade, Serbia

Abstract

The increased demand for cheap, biodegradable, renewable, and recyclable fibers with exceptional properties positioned jute (*Corchorus capsularis* L. and *Corchorus olitorius* L.) in the second place (after cotton) in the natural fiber world market. Multicellular jute fibers are comprised of three main components: cellulose, lignin, and hemicelluloses having a wide variety of functional groups capable of binding different water pollutants. However, such groups are not easily accessible due to the presence of a hydrophobic surface layer (consisting of pectins, waxes, and fats) that could be removed by applying simple alkali and oxidative modifications. Moreover, fibers' activation using different chemical agents or grafting of functional groups on their surfaces results in enhanced fiber sorption properties, and hence adsorption potential for various water pollutants. This chapter provides an overview of the possibility of the application of raw and chemically modified jute fibers as an eco-friendly adsorbent for heavy metals and dyes as the most frequent water pollutants. Special attention has been paid to the binding mechanism of the pollutants and differently functionalized jute adsorbents. The last section of this chapter represents

---

[*] Corresponding Author's Email: aivanovska@tmf.bg.ac.rs.

In: Jute: Cultivation, Properties and Uses
Editor: Matthieu Issa
ISBN: 979-8-88697-490-4
© 2023 Nova Science Publishers, Inc.

one step toward both the circular economy approach and sustainable development, in terms of reusing and revalorization of solid waste with adsorbed pollutants. Permanent collection and reuse of pollutant saturated jute adsorbents have promising multi-positive effects on the economy as well environment, including reducing its quantity, saving energy, and its utilization as raw material for producing new hybrid materials which is in line with the Circular Economy Package (2020).

Keywords: jute, chemical modifications, sorption properties, adsorbent, pollutants, heavy metals, dyes

Introduction

Besides bringing material convenience, industrial civilization and modernization caused uncontrolled environmental pollution that represents a potential threat to the well-being of life on earth (Abarna et al., 2016). The majority of wastewater is neither collected nor treated and often enters water sources affecting human health, cutting down the availability of hygienic foods and drinks, and giving rise to various epidemic diseases (Huang et al., 2019). The main constituents of wastewaters are the domestic, food industry, tannery, textile, pharmaceutical, medical, mining, and distillery wastewaters (Perumal et al., 2022). Therefore, the treatment of wastewaters emerges as one of the fundamental problems of today and a constant effort must be made to protect water resources. Many different wastewater treatments such as physical (boiling, distillation, filtration, and sedimentation), chemical and physicochemical (adsorption, ion exchange, reverse osmosis, chemical oxidation, ultraviolet, photocatalytic and sonochemical degradation), and biological (microbial water sludge treatment) and their combinations were used to remove insoluble particles and soluble contaminants from wastewaters. Some of them such as ion exchange and reverse osmosis, and advanced oxidation processes are too expensive to build, operate, and maintain, especially in developing countries (Roy, 2021). On the other hand, chemical precipitation, membrane filtration, ion exchange, and reverse osmosis are ineffective at low pollutant concentrations (i.e., below 100 mg/l), generate a toxic sludge (resulting in a new kind of pollution), use high quantities of chemicals, and involve high energy inputs (Crini and Lichtfouse, 2019; Ivanovska et al., 2022a).

Nowadays, the adsorption process has received more and more attention since it can overcome the mentioned shortcomings of other wastewater treatments, especially the high cost of wastewater treatment. More precisely, the advantages of adsorption concerning the aforementioned treatments are multiple: (1) simplicity, flexibility, and ease of implementation, (2) adsorbents are in many cases cheap and easily available and can be selective, (3) adsorption can be used for the treatment of wastewater in which the concentration of pollutants is less than 100 mg/l, (4) high removal efficiency, (5) no secondary pollutants are created, (6) possibility of regeneration and/or revalorization of the adsorbent (Loiacono et al., 2018; Chen et al., 2020; Ivanovska et al., 2020a; Sahu, Mahapatra and Patel, 2017). To be a truly economical and efficient process, it is necessary to choose or synthesize a suitable adsorbent that has the lowest possible price and the highest affinity for the given pollutant.

With the depletion of global oil and petrochemical resources and the growing interest in environmental issues, the pursuit of using renewable and sustainable resources is increasingly attracting the attention of both the scientific community and various industries. One of the most effective strategies for sustainability is based on the use of adsorbents obtained from natural sources, among which lignocellulosic materials are the most represented and almost inexhaustible ones. Various lignocellulosic materials, e.g., raspberry canes (Kukic et al., 2022), hemp shives (Mongiovi et al., 2022a), rice husk (Mladenovic et al., 2020), wood-based materials (Khera et al., 2019; Ivanovska et al., 2021a), potato peels (Farooq et al., 2019), etc. have been already used as adsorbents. Among them, the jute (*Corchorus capsularis* L. and *Corchorus olitorius* L.) being an abundant (took second place behind cotton in the natural fiber world market (Ivanovska et al., 2022b) with a total production of 2.69 million t in 2020 (FAO, 2021)) and inexpensive fiber offered particular attraction as an adsorbent.

Jute fibers are multicellular (comprised of 6-20 elementary fibers) and are recognized by their heterogeneous chemical composition that includes cellulose (58-63%) and noncellulosic components like lignin (11.4-12.0), and hemicelluloses (21.0-24.0%) having a myriad of functional groups (Ivanovska et al., 2019) capable of binding different pollutants. Such functional groups are not easily accessible due to the existence of a hydrophobic surface layer of pectins, waxes, and fats that could be successfully removed by applying simple alkali and oxidative modifications. Furthermore, the most promising strategies for fiber functionalization and increasing their potential for adsorption of various pollutants are their activation using different chemical agents or grafting of functional groups on the fibers' surfaces.

The following literature review gives a more detailed insight into the possibility of application of jute fibers as an eco-friendly adsorbent for the most frequent pollutants (i.e., heavy metals and dyes), which is very important from aspects of sustainable development and environmental protection. The last section of this chapter represents one step toward both the circular economy approach and sustainable development, in terms of management and revalorization of solid waste with adsorbed pollutants.

Adsorption of Heavy Metal Ions by Raw and Modified Jute

The increased contamination of water sources with heavy metals such as lead, zinc, copper, mercury, arsenic, chromium, nickel, and cadmium represents a growing concern for people from all over the world (Anderson et al., 2022; Chakraborty et al., 2022). It is well known that heavy metals could originate from both natural and anthropogenic sources. Natural sources mainly comprise bedrock weathering, while anthropogenic sources include various industries (primarily electroplating, leather tanning, dyeing, and mining industry), coal combustion, unplanned urbanization, solid waste disposal, and agricultural soil fertilization (Zhou et al., 2020; Kumar et al., 2019; Kukic et al., 2022). Heavy metals are considered as the most dangerous inorganic water pollutants since they are non-biodegradable; once released into the environment, they remain there permanently and accumulate in concentrations higher than permitted. Their presence in aqueous systems at trace levels has become a worldwide health concern due to the possibility of entering the human body through water, food, and air (Meng, Bai and Tang, 2022), accumulating in the tissues and thus causing neurological diseases, anemia, kidney dysfunction, cancer, etc. The relationship between long-term exposure to heavy metals and various human diseases was investigated by Buha et al. (2021), Das et al. (2018), Nurchi et al. (2020), Bjørklund et al. (2019), and many other researchers.

Adsorption of heavy metals onto lignocellulosic fibers proved to be an adequate method for minimizing their concentrations in wastewaters and therefore prevents the mentioned human diseases. Having in mind that the jute fibers' main chemical components (i.e., cellulose, hemicelluloses, and lignin), as well as other minor components, have various functional groups responsible for binding metal ions, they were evaluated as adsorbents for heavy metals present in wastewaters (Ivanovska et al., 2020b). Strong bonding of metal ions by carboxylic (primarily present in hemicelluloses, pectin, and lignin), phenolic (lignin and extractives), and to a certain extent aldehyde (lignin) and hydroxyl groups (cellulose, hemicelluloses, lignin, extractives, and pectin) groups often involves complexation (donation of an electron pair from these groups to form complexes with the metal ions in solution) and ion exchange, cation-π interactions, and electrostatic interactions (Ivanovska et al., 2020a; 2021a). Moreover, the metal adsorption through physical forces and "trapping of ions" in fiber interfibrillar and intrafibrillar cracks and cavities should not be neglected.

To enhance the jute fiber adsorption potential for heavy metals, chemical modifications using alkalies such as NaOH, oxidative agents like $NaClO_2$, H_2O_2, NaOCl, and other organic (dyes, polyaniline) and inorganic (iron oxyhydroxide) agents were investigated, Table 1. Additionally, coatings with nanoparticles and grafting procedures were also employed to increase the fiber adsorption potential. In the following text, the adsorption potential of chemically modified jute will be discussed in detail.

Far away, Shukla and Sakhardande (Shukla and Sakhardande, 1991a) used organic dyes C.I. Reactive Red 31, C.I. Reactive Orange 13, and C.I. Reactive Yellow 18 for enhancing the jute fiber adsorption potential for Fe^{2+}, Fe^{3+}, Pb^{2+}, and Hg^{2+} from $FeSO_4 \cdot 7H_2O$, $FeCl_3$, $(CH_3-COO)_2Pb \cdot 3H_2O$, and $HgCl_2$ solutions. Based on the conducted batch experiments, i.e., decline in metal solution pH after the adsorption, the authors concluded that jute acts as an acidic ion exchanger. Both raw and dyed jute fabrics possessed the highest adsorption potential for Fe^{3+}; the dyeing improved fiber adsorption potential for the studied metals up to 6.3 times, Table 1. The authors broadened their investigations and used jute fibers dyed with C.I. Reactive Red 31 (Shukla and Sakhardande, 1991b) and C. I. Reactive Orange 13 (Shukla and Sakhardande, 1992) as adsorbents for Pb^{2+}, Hg^{2+}, Cu^{2+}, Fe^{3+}, Zn^{2+}, Ni^{2+} and Fe^{2+} in the semi-continuous system using packed columns. The performed experiments proved that in the case of high initial metal concentrations (1200-1300 mg/l), the column adsorption is more effective than the batch adsorption. Moreover, the columns could be regenerated using diluted acids and used repeatedly for adsorbing different metals several times.

In the literature data (Shukla and Pai, 2005a), the adsorption of Cu^{2+}, Ni^{2+}, and Zn^{2+} from monometallic aqueous solutions onto raw and jute fibers modified with H_2O_2 or C. I. Reactive Orange 13 was studied. The hydrogen peroxide modification aimed to oxidize the cellulose hydroxyl to carboxyl groups, and hence obtain a weak cation exchanger. After both applied modifications, the jute fibers possessed higher adsorption potentials for metal ions than before them. The oxidized and dyed jute possessed the highest affinity towards Zn^{2+} and Cu^{2+} (~ 8.0 and 8.4 mg/g, respectively), Table 1. Equilibrium adsorption data showed that the adsorption of studied ions best fit the Langmuir isotherm model. Additionally, the desorption efficiency, regenerative, and reuse capacity of these adsorbents were assessed for three successive adsorption-desorption cycles, and the efficiency for reuse was maintained when after the desorption, the fibers were treated with 0.5 g/l NaOH for 60 min. The adsorption capacities of oxidized fibers for Cu^{2+}, Ni^{2+}, and Zn^{2+} after three cycles of successive adsorption-desorption accounted for 53.14, 56.43, and 45.57%, while in the case of dyed fibers these percentages were 81.20, 77.73 and 65.06% compared to the values achieved after the first adsorption cycle (Shukla and Pai, 2005a). The same authors (Shukla and Pai, 2005b) compared the adsorption capacities of raw and C. I. Reactive Orange 13 modified jute fibers for Pb^{2+}. However, the improvement of Pb^{2+} adsorption onto dyed jute fibers was not as high as in the case of the adsorption of Cu^{2+}, Ni^{2+}, and Zn^{2+}, Table 1 (Shukla and Pai, 2005a). This behavior could be explained by more than three times higher molecular weight and about 1.5-2.0 times higher ionic radius, as well as the lower electronegativity of Pb^{2+}, compared to the other three studied metals (Loiacono et al., 2018).

The last statement could be also used for the explanation of the different affinity of Ni^{2+}, Cu^{2+}, and Zn^{2+} for jute fabrics' active sites in the case of competitive, i.e., polymetallic adsorption (Ivanovska et al., 2020a). Namely, the authors underlined that the larger the electric charge density is, the lower the molar mass and effective ionic radii are, the higher the ion affinity is. Moreover, the influence of alkali and oxidatively modified jute fabrics' chemical compositions, the content of functional groups, and experimental conditions on the adsorption of Ni^{2+}, Cu^{2+}, and Zn^{2+} from monometallic solutions was investigated, (Ivanovska et al., 2020a; 2020b; 2021b). The maximum adsorption capacity for all studied heavy metal ions was observed at a pH of 5.50. The raw jute attained adsorption equilibrium faster than fabrics modified with NaOH or $NaClO_2$. Increased initial metal ion concentration from 10 to 20 mg/l caused an increase in the total uptake capacity of chemically modified jute fabrics for up to 69, 63, and 37% for Ni^{2+}, Cu^{2+}, and Zn^{2+} (Figure 1, Ivanovska et al., 2020a), respectively.

Table 1. Comparison between the jute adsorption capacities for different metal ions

Adsorbent	Metal	Adsorbent dose, g/l	pH	Contact time, min	Temperature, °C	c_0, mg/l	q_{ex}, mg/l	Reference
Raw jute fibers	Fe^{2+}	20	N/A	120	N/A	N/A	10.1	Shukla and Sakhardande (1991a)
	Fe^{3+}						18.8	
	Pb^{2+}						5.0	
	Hg^{2+}						3.2	
Jute fibers dyed with C.I. Reactive Red 31	Fe^{2+}					120	15.6	
	Fe^{3+}						25.9	
	Pb^{2+}					130	18.8	
	Hg^{2+}						19.0	
Jute fibers dyed with C.I. Reactive Orange 13	Fe^{2+}					120	15.6	
	Fe^{3+}						26.9	
	Pb^{2+}					130	19.5	
	Hg^{2+}						18.2	
Raw jute fibers	Cu^{2+}	20	5.5	120	35	N/A	4.2*	Shukla and Pai (2005a)
	Ni^{2+}		6.6				3.4*	
	Zn^{2+}		5.9				3.6*	
Jute fibers modified with H_2O_2	Cu^{2+}		5.5				7.7*	
	Ni^{2+}		6.6				5.6*	
	Zn^{2+}		5.9				8.0*	
Jute fibers dyed with C.I. Reactive Orange 13	Cu^{2+}		5.5				8.4*	
	Ni^{2+}		6.6				5.3*	
	Zn^{2+}		5.5				6.0*	
Raw jute fibers	Pb^{2+}	20	4.9	120	25	N/A	16.0	Shukla and Pai (2005b)
Jute fibers dyed with C.I. Reactive Orange 13							18.6	

Table 1. (Continued)

Adsorbent	Metal	Adsorbent dose, g/l	pH	Contact time, min	Temperature, °C	c_0, mg/l	q_{ex}, mg/l	Reference
Raw jute fabric	Ni^{2+}	2.5	5.5	420	25	20	3.8	Ivanovska et al. (2020a; 2020b; 2021a)
	Cu^{2+}						2.4	
	Zn^{2+}						2.8	
Jute fabric modified with NaOH	Ni^{2+}						5.0	
	Cu^{2+}						4.4	
	Zn^{2+}						3.8	
Jute fabric modified with NaClO$_2$	Ni^{2+}						6.1	
	Cu^{2+}						6.1	
	Zn^{2+}						4.8	
Raw jute fibers	Cu^{2+}	2	5.5	15	28	100	~6.0	De Quadros Melo et al. (2015)
	Ni^{2+}						~11.0	
	Pb^{2+}						~4.0	
	Cd^{2+}						~4.0	
Jute fibers modified with NaOH	Cu^{2+}						~8.0	
	Ni^{2+}						~12.5	
	Pb^{2+}						~13.0	
	Cd^{2+}						~8.0	
Jute fibers modified with polyaniline	Cr^{6+}	2	3.0	150	20	100	~15.0*	Kumar, Chakraborty and Ray (2008)
Jute fibers grafted with AA[a]	Hg^{2+}	4	6.0	60	30	27.15	6.8	Hassan and Zohdy (2018)
	Pb^{2+}		5.0			33.15	7.2	
Jute fibers grafted with PMDA[b]	Pb^{2+}	1	6.0	60	25	200	157.2	Du et al. (2016)
	Cd^{2+}						88.0	
	Cu^{2+}						44.0	

Adsorbent	Metal	Adsorbent dose, g/l	pH	Contact time, min	Temperature, °C	c_0, mg/l	q_{ex}, mg/l	Reference
Jute fibers coated with Fe_2O_3 nanoparticles	As^{5+}	0.2	3.0	150	25	50	~35.0	Sahu, Mahapatra and Patel (2017)
Jute fibers loaded with iron oxyhydroxide	As^{3+}	N/A	7.0	200	25	N/A	12.7*	Hao et al. (2015)
Raw jute yarns	Cu^{2+}	10	N/A	60	N/A	100	0.85	Al-Mamun et al. (2010)
AA^a grafted jute yarns							1.25	
AA^a and PA^c grafted jute yarns							4.78	
Jute nanofibers successively modified with NaOH and NaOCl	Hg^{2+}	1	6.0	60	25	N/A	31.0	Baheti et al. (2013)

N/A – not applicable.
* – Maximal adsorption potential obtained from isotherm experiments.
a – Acrylic acid, b - 1,2,4,5-Benzenetetracarboxylic anhydride, c – phosphoric acid.

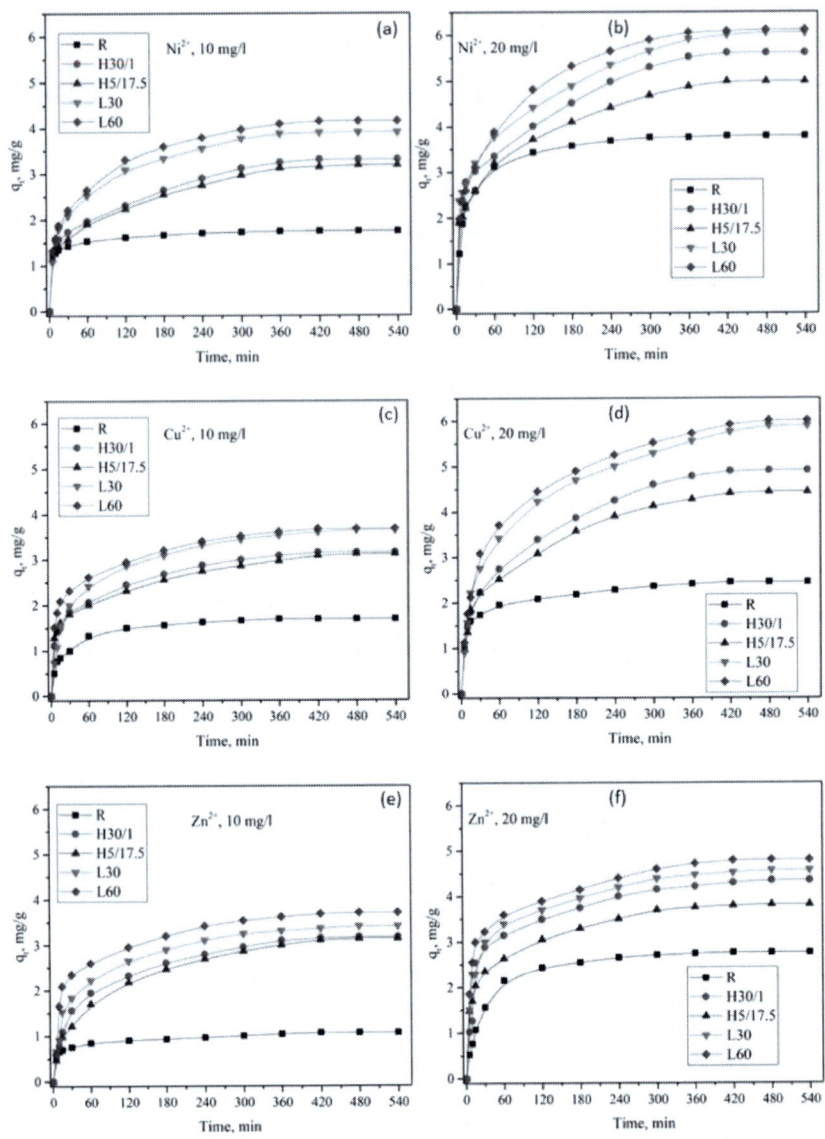

Figure 1. Adsorption capacities (q, mg/g) of raw (R), alkali (H30/1, and H5/17.5), and oxidized (L30, and L60) jute fabrics for Ni^{2+}, Cu^{2+}, and Zn^{2+} from the solution with initial ion concentrations of 10, and 20 mg/l (Ivanovska et al., 2020a) (Reprinted by permission from Springer Nature Customer Service Centre GmbH: Springer Nature: *Fibers and Polymers,* Waste jute fabric as a biosorbent for heavy metal ions from aqueous solution, Ivanovska A, Dojcinovic B, Maletic S, Pavun L, Asanovic K, Kostic M., Copyright (2020)).

The best adsorption performance possessed jute fabric oxidized with 0.7% $NaClO_2$ for 60 min (having 63.2% lower lignin content and 81.1% higher amount of carboxyl groups compared to the raw jute). Its adsorption capacity for Ni^{2+}, Cu^{2+}, and Zn^{2+} was about 2.4; 2.2, and 3.5 times higher than those of the raw jute fabric, Table 1. Further, the kinetic adsorption experiments (Ivanovska et al., 2021b) revealed that the adsorption of nickel ions onto all studied fabrics followed the pseudo-second-order kinetic model, while the experimental isotherm data fit with the Langmuir model supporting the assumption that the adsorption process is monomolecular and occurred at fabrics' homogenous active sites. The calculated ratios between fabric maximal adsorption capacity and carboxyl group content indicated that approximately 1/3 of the fabrics' carboxyl groups would be involved in binding nickel ions during adsorption, Figure 2.

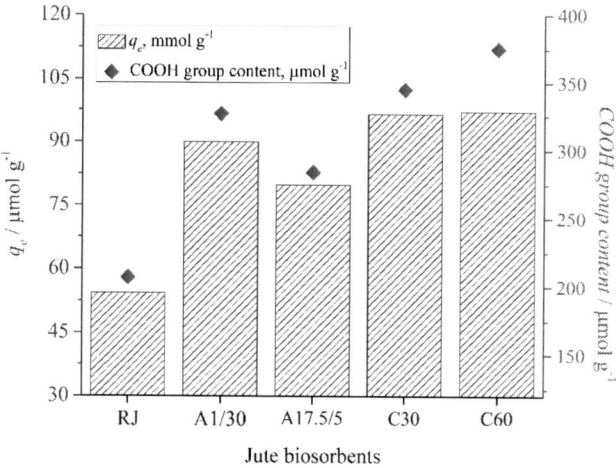

Figure 2. Equilibrium adsorption capacities of raw (RJ), alkali (A1/30 and A17.5/5), and oxidatively (C30 and C60) modified jute fabrics for nickel ions *vs.* COOH group content (Ivanovska et al., 2021b), CC BY 3.0 (https://creativecommons.org/licenses/by/3.0/)

Besides the fact that the content of carboxyl groups could be used to predict the maximal adsorption capacity of jute fabrics toward nickel ions, the correlation between these two variables was also confirmed for the adsorption of Zn^{2+} onto the raw (J), alkali (JA5/17.5) and oxidatively (JO60) modified jute fabrics (Ivanovska et al., 2020b). Namely, after the alkali modification, the Zn^{2+} adsorption (initial metal concentration of 15 mg/l) was improved 2.7 times, which can be attributed to the 27.1% higher amount of carboxyl groups as well as 36.6% lower hemicellulose content, Figure 3. Moreover, the oxidized jute fabric possessed the highest adsorption potential for the same metal since the oxidation leads to the removal of lignin and other non-cellulosic components, and therefore increases the available hydrophilic fiber surface area. In parallel, a higher amount of carboxyl groups was observed (0.375 mmol/g, Figure 3) due to the sodium chlorite oxidation of residual lignin.

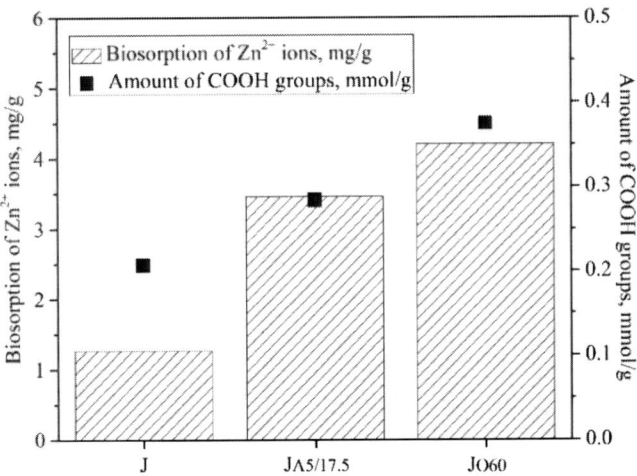

Figure 3. a) Adsorption of Zn^{2+} (c_0 = 15 mg/l) *vs.* the amount of COOH groups in the jute fabrics (Ivanovska et al., 2020b) (Reprinted by permission from Springer Nature Customer Service Centre GmbH: Springer Nature: *Cellulose,* Multifunctional jute fabrics obtained by different chemical modifications, Ivanovska A, Asanovic K, Jankoska M, Mihajlovski K, Pavun L, Kostic M., Copyright (2020)).

The modeling of the isotherm data pointed out that adsorption/binding of Zn^{2+} is mediated by chemical rather than physical forces of attraction, which was additionally confirmed by the ATR-FTIR spectra (i.e., decreasing of the peaks' maxima after the adsorption of Zn^{2+}, Figure 4). The discussed paper (Ivanovska et al., 2020b) is very interesting not only from the aspect of metal adsorption but also for further utilization of jute fabrics as filters for water disinfection. Namely, fabrics obtained after Zn^{2+} adsorption provided maximum bacterial reduction (99.99%) for *Escherichia coli* and *Staphylococcus aureus* enabling creation of the closed technological loop focused on the reduction of lignocellulosic textile waste and strengthening of recycling processes.

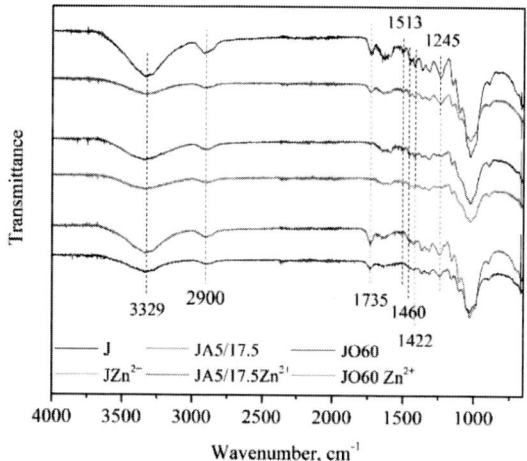

Figure 4. ATR-FTIR spectra recorded before and after adsorption of Zn^{2+} onto jute fabrics (Ivanovska et al., 2020b) (Reprinted by permission from Springer Nature Customer Service Centre GmbH: Springer Nature: *Cellulose,* Multifunctional jute fabrics obtained by different chemical modifications, Ivanovska A, Asanovic K, Jankoska M, Mihajlovski K, Pavun L, Kostic M., Copyright (2020)).

In the previously discussed papers, jute was alkali modified in the form of fabric, while a group of authors (De Quadros Melo et al., 2015) investigated how the different alkali modifications (5, 7, or 10% NaOH) of jute in the form of fibers affected their adsorption potential for Cu^{2+}, Ni^{2+}, Pb^{2+}, and Cd^{2+} from polymetallic solution, Table 1. According to the obtained results, the highest adsorption potential possessed jute fibers modified with 10% NaOH. The content of adsorbed heavy metal ions by alkali modified jute fibers was presented in the following order: Pb^{2+}> Ni^{2+}> Cu^{2+} ~ Cd^{2+}. The equilibrium adsorption was attained after 15 min of contact time, while the isotherm experimental data were in good agreement with the Langmuir model (De Quadros Melo et al., 2015).

Although alkali and oxidatively modified jute was successfully used as adsorbents for various metal ions, the fiber surface was also modified by the synthesis of short-chain polyaniline (PANI). Such prepared fibers were used as an adsorbent for Cr^{6+} (Kumar, Chakraborty and Ray, 2008). Complex adsorption experiments by varying the solution pH, initial Cr^{6+} concentration, adsorbent dose, and temperature were carried out. The obtained results showed that maximum monolayer adsorption of total chromium (Cr^{6+}, Cr^{3+}, and its other forms) of 62.9 mg/g was observed at a solution pH of 3 and temperature of 20°C. When the Cr^{6+} initial concentration was 100 mg/l, the adsorption equilibrium was achieved after 150 min, Table 1. Total chromium adsorption followed Lagergren's pseudo-second-order and Elovich models indicating that its adsorption is of chemical nature. Raising the adsorption temperature resulted in a decrease in the adsorption potential of modified jute suggesting exothermic adsorption. The desorption efficiency of 83% was achieved for 10 min in 2 M NaOH, however, upon ignition of PANI-jute-chromium, almost 94% recovery was obtained as Cr^{3+} ion (Kumar, Chakraborty and Ray, 2008). In further investigations, Kumar and Chakraborty (2009) conducted fixed-bed column studies to evaluate the adsorption performance of polyaniline-modified jute fibers for Cr^{6+} from an aqueous solution. Based on the obtained results, the following conclusions were drawn: (1) pH of 3 was optimal for the adsorption of total chromium (i.e., Cr^{6+} and Cr^{3+}) of 26.74 mg/g at a flow rate of 2 ml/min and initial Cr^{6+} concentration of 20 mg/l; (2) total chromium removal was well described by BDST equation till 10% breakthrough, and (3) adsorption rate constant and dynamic bed capacity at 10% breakthrough were observed as 0.01 l/mg h and 1069.46 mg/l, respectively. All the adsorbed total chromium was recovered back from polyaniline-modified jute fibers as non-toxic Cr^{3+} after ignition with more than 97% reduction in weight.

Grafting is another method that was used for increasing the jute fiber adsorption potential for heavy metal ions. In the research conducted by Hassan and Zohdy (2018), previously swelled (treated with 20% NaOH for 24 h) jute fibers were grafted with acrylic acid by using the direct gamma irradiation technique. After that, the adsorption of toxic heavy metals Hg^{2+}, and Pb^{2+} from their aqueous solutions onto prepared adsorbents was studied. The adsorption process obeys the second-order kinetic reaction and follows the Langmuir adsorption isotherm model, while, about 86 and 80% of Pb^{2+} and Hg^{2+} were adsorbed after a contact time of 60 min. Somewhat higher adsorption potential for Pb^{2+} than for Hg^{2+} (7.2 vs. 6.8 mg/g) lies in the smaller hydrated ionic radii of Pb^{2+}, which increase its adsorption possibility onto grafted jute fibers.

Another study in which grafted jute fibers were used as adsorbents for Pb^{2+}, Cd^{2+}, and Cu^{2+} was reported by Du et al. (2016). Namely, jute fibers were preswelled (using 15% NaOH for 10 min at 368 K) in a microwave reactor and then grafted with 1,2,4,5-Benzenetetracarboxylic anhydride (PMDA) in N,N-Dimethyl formamide solution for about 15 min at 393 K in the same reactor. The highest adsorption potential was observed for Pb^{2+}, while the lowest for Cu^{2+}, Table 1. The adsorption of studied metals onto carboxyl-modified jute fibers followed Langmuir isotherm model. On the other hand, the adsorption equilibrium was achieved within 20 min of contact time and follows the pseudo-second-order kinetic model. Based on the thermodynamic studies, the adsorption was found to be spontaneous and endothermic. The enhanced metal adsorption onto grafted jute fibers may be attributed to the introduction of -COOH groups on their surfaces, and the main mechanism of metal adsorption was the ion exchange. The last one was confirmed by the small change of solution electrical conductivity after adsorption. Grafted fibers with adsorbed metals can be easily regenerated with ethylenediaminetetraacetic acid disodium salt solution and reused up to at least four times with equivalent high adsorption capacity. In further experiments conducted by Du, Zheng and Wang (2018), the preswelled jute fibers were grafted with Pyromellitic dianhydride in N,N-Dimethyl formamide solution for about 15 min at 393 K. The purpose of the described grafting procedure was to introduce metal-binding groups (-COOH) on the jute surface. About ten times higher adsorption of Cu^{2+} onto carboxyl-modified jute fibers than onto raw jute (43.98 vs. 4.23 mg/g) proved that they were successfully introduced by grafting. This paper differs from the previously published one since the adsorption was performed in a fixed-bed column and the jute fibers were grafter with a different agent.

Ramim et al. (2017) employed different methods for coating raw and bleached jute fabrics with iron oxide/hydroxides. After that, the adsorption of As^{3+} and As^{5+} onto the prepared fabrics having different structural characteristics (i.e., mass per unit area, and thickness) was studied. However, the results are not comparable with those reported in other studies since they are presented in mg/m^2. On the other hand, As^{5+} was successfully adsorbed onto jute fibers coated with Fe_2O_3 nanoparticles (Sahu, Mahapatra and Patel, 2017). Different instrumental methods and techniques (such as XRD, FTIR, elemental analysis, SEM, and TEM) were used to confirm that Fe_2O_3 nanoparticles are present on the jute fiber surface. The prepared nanocomposite showed the adsorption potential of about 35 mg/g (initial metal concentration of 50 mg/l) at a pH of 3 and room temperature, Table 1. The maximum adsorption capacity calculated from the Langmuir isotherm model was found to be 48.06 mg/g. The As^{5+} adsorption mechanism onto jute-Fe_2O_3 nanocomposite involved both electrostatic attraction and surface complexion including bidentate and monodentate ligand exchange mechanism. The effect of other coexisting oxyanions on the As^{5+} adsorption onto nanocomposite was presented in the following order: PO_4^{3-} > SiO_3^{2-} > CO_3^{2-} > SO_4^{2-} > NO_3^-. Additionally, the desorption experiments were done using 0.5 M NaOH solution and the nanocomposite could be regenerated up to four times.

Compared to the discussed studies, Hao et al. (2015) have conducted more complex experiments in which the adsorption of As^{3+} onto jute fibers with loaded iron oxyhydroxide (hybrid Fe-jute fibers) was investigated. Briefly, jute fibers were firstly esterified with succinic anhydride under pyridine reflux at 90°C for 12 h. After the subsequent washing with acetone, anhydrous ethanol, and deionized water, the introduction of Fe^{3+} was carried out by immersing the carboxylate-functionalized jute fibers in an aqueous solution of 0.05 M $Fe(NO_3)_3 \cdot 9H_2O$ (pH of 7.0) for 12 h. After filtration, the fibers were filled into a glass column, and then 0.5 mol/l NaOH was slowly dropped for 2 h. The reaction takes place for another 12 h. The X-ray diffraction pattern indicated that the loaded iron oxyhydroxide was mainly α-FeOOH. Thereafter, the Fe-jute fibers were used as adsorbents for As^{3+} and the maximum adsorption potential of 12.66 mg/g (at a pH of 7.0) was determined based on the Langmuir isotherm model, Table 1. Flow rate has a significant influence on the removal rate; the saturation concentration simulated by the Adams-Bohart model was found to be 159.21 mg/l at an empty-bed contact time of 3.5 min. Moreover, the leaching of iron during the adsorption (0.178 mg/l) was found to be lower than the standard limit. The coexisting anions such as Cl^-, SO_4^{2-}, NO_3^-, F^-, and SiO_3^{2-} did not affect the As^{3+} adsorption.

In the already discussed papers, jute fibers or fabrics were used as adsorbents for heavy metal ions. However, jute yarns were also evaluated as the selective ion adsorbents after curing with acrylic acid (AA) and phosphoric acid (PA) using UV radiation (Al-Mamun et al., 2010). The raw, as well as yarns grafted with AA and with both AA and PA, were used as adsorbents for Cu^{2+} present in an aqueous solution at a trace level, i.e., an initial concentration of 10 mg/l. According to the equilibrium adsorptions of 0.85, 1.25, and 4.75 mg/g (Table 1) observed for raw, jute yarns grafted with AA and those grafted with AA and PA, respectively, it is clear that grafting with AA and PA significantly improved the adsorption potential of jute yarns. In the same study (Al-Mamun et al., 2010), the authors reported that 1.45 mg of copper were recovered per gram of yarns in 0.04 M HCl for 3 h.

There is only one literature data (Baheti et al., 2013) in which jute in the form of nanofibers was used as an adsorbent for metal, i.e., Hg^{2+}. In order to obtain nanofibers, raw jute was successively treated with 4% NaOH at 80°C for 1 h, and 7 g/l NaOCl at room temperature for 2 h at a pH of 10-11. The adsorption of Hg^{2+} was optimized in terms of solution pH (2-10), initial metal concentration (100-500 mg/l), adsorbent dose (0.1-5.0 g/l), contact time (5-120 min) and temperature (25-55 °C). The maximal adsorption capacity was observed at a pH of 6.00, while the adsorption equilibrium was reached after 60 min. The obtained results revealed that with increasing the adsorbent dose, the jute fibers' adsorption potential increases too. Adsorption efficiency for

Hg^{2+} increased from 31.5 to 95.6%, while the adsorption capacity decreased from 31.0 down to 9.6 mg/g when the adsorbent dose was increased 10 times (from 0.1 to 1.0 g/l). Furthermore, the fibers' adsorption potential decreased from 95.6 down to 80.5% when the temperature raised from 25 to 55°C (Baheti et al., 2013).

Lignocellulosic fibers such as jute are appropriate carbon precursors since they are comprised of carbon-rich components such as cellulose, hemicelluloses, and lignin (Mongioví et al., 2022b). The specific surface area, amount of surface oxygen groups, and morphology of resulting carbons are strongly affected by the precursor chemical composition and structure as well as the applied activation processes. Fibrous activated carbons prepared from jute (as a natural precursor) by physical and/or chemical activations were used as adsorbents for Cu^{2+} (Phan et al., 2006), Cr^{3+}, and Cd^{2+} (Hossain et al., 2019), and Cr^{6+} (Chen et al., 2020). The obtained adsorption potentials for Cr^{3+} and Cd^{2+} are 16.6% and 12.9% higher in comparison with those of commercially available activated carbon (Hossain et al., 2019). Further, adsorption experiments demonstrated that 98% of Cr^{6+} present in water could be removed by biochar under a wider operating pH range (1-6), whereby Cr^{6+} was reduced to Cr^{3+} with a reduction percentage between 86 and 97% (Chen et al., 2020).

Jute, a Promising Adsorbent for Synthetic Dyes

Synthetic dyes represent very complex organic pollutants that are present in the wastewaters of rapidly developed industries such as textile, cosmetics, paper, leather, rubber, wood, plastic, printing, etc. Unfortunately, colored wastewaters are often released into the streams and rivers that are located near the industries (Fodeke and Olayera, 2019). Independently on dye structure and even when the dyes are present in low concentrations, such wastewaters cause environmental pollution. Namely, the degradation products of many synthetic dyes are toxic and mutagenic and represent aquatic ecology disruptors (Calimli et al., 2020; Han et al., 2022). Besides color, wastewaters contain solids and salts and have increased biochemical and chemical oxygen demand (BOD and COD, respectively), pH, and temperature (Yaseen and Scholz, 2019; Laizer et al., 2022). Above-discussed encouraged many researchers to focus on the decoloration of wastewaters through adsorption onto different cellulosic materials such as jute in the form of fibers, fabrics, or yarns, Table 2.

Table 2. Comparison between the jute adsorption capacities for different dyes

Adsorbent	Dye	Adsorbent dose, g/l	pH	Contact time, min	Temperature, °C	c_0, mg/l	q_{ex}, mg/l	Reference
Raw jute fibers	Congo Red	4	7.8	20	30	50	~7.0	Banerjee and Dastidar (2005)
	Methylene Blue		6.5	30			~9.8	
	Acid Violet		7.0	60			~5.0	
Raw jute yarns	Acry Red 4G	20	4.0	120	25	N/A	1.97*	Prajapati et al. (2005)
Raw jute fibers	Congo Red	2	4.0	180	30	50	~2.8	Roy et al. (2013a)
Tanin-modified jute fibers	Congo Red	6	3.0	30	30	50	~7.5	Roy, Adhikari and Majumder (2013b)
Raw jute fabric	Congo Red	5	10.0	330	25	50	4.7	Ivanovska et al. (2022a)
Jute fabric oxidized with NaIO$_4$							8.5	
Jute fibers modified with NaOH	Congo Red	14.52	7.2	N/A	30	150	33.7	Dey, Kumar and Dey (2018)
	Reactive Red 195	14.90	7.0	120	26.4	50	32.2	Dey and Dey (2021)
Raw jute fabric	C.I. Acid Blue 111	2.5	3.5	300	25	N/A	12.9*	Ivanovska et al. (2021c)
Jute fabric modified with NaOH							16.4*	
Jute fabric modified with NaClO$_2$							17.0*	
Jute fibers modified with pyromellitic dianhydride	Aniline	3	7.30	240	25	50	30.7	Gao et al. (2015)
Jute fibers grafted with acrylic acid and immobilized with chitosan	Reactive Blue 222	N/A	7.0	120	25	60	220	Hassan (2015)

N/A – not applicable.
* – maximal adsorption potential obtained from isotherm experiments.

The oldest data regarding the utilization of jute fibers as adsorbents date back to 1946 (Sarkar, Chatterjee and Mazumdar, 1946) and 1948 (Sarkar and Chatterjee, 1948). In the first paper, the adsorption of Methylene Blue onto defatted jute fibers was carried out at room temperature for 2 h and an adsorption capacity of 16.72 mg/g was obtained (Sarkar, Chatterjee and Mazumdar, 1946). Two years later, the experiments were extended and the authors Sarkar and Chatterjee (1948) studied the adsorption potential of raw, defeated, alkali-treated, acid-treated, and oxidized jute fibers for Methylene Blue under various experimental conditions. Among all studied adsorbents, defeated jute possessed the lowest, while oxidized jute possessed the highest adsorption potential (41.9 vs. 87.3 mg/g).

After a gap of more than a half-century, this topic becomes relevant again and Banerjee and Dastidar (2005) directed their research on the adsorption potential of jute fibers for three different synthetic dyes. The listed results (Table 2) revealed the following fibers' adsorption potential order: Methylene Blue (9.8 mg/g) > Congo Red (7.0 mg/g) > Acid Violet (5.0 mg/g). As is evident from Table 2, besides Methylene Blue, azo dye Congo Red was chosen as a model dye in many studies, since, from one side, it can not readily degrade under neutral conditions, while from the other side, it is suitable for studying its adsorption mechanism with jute adsorbent (Ivanovska et al., 2022a).

It has to be emphasized that there is only one available literature in which jute was used as an adsorbent in the form of yarn (Prajapati et al., 2005). The experimental data for the adsorption of Acry Red 4G onto jute yarn fit well with the pseudo-second-order model indicating that the chemical sorption is the rate-limiting step.

Besides batch experiments, Roy et al. (2013a) also conducted adsorption of Congo Red onto jute fibers in fixed bed column conditions. In the case of the batch studies, the dye uptake was highly dependent on initial pH and dye concentration, adsorbent dosage, contact time, ionic strength, and temperature. The adsorption was characterized as exothermic and spontaneous; it agreed well with the Langmuir isotherm and pseudo-second-order kinetic model. When the column studies were conducted, the obtained results indicated that the total amount of adsorbed dye decreased with increasing flow rate and increased with increasing bed height and initial dye concentration. Also, the breakthrough and exhaustion time increased with increasing bed depth but decreased with increasing flow rate and dye concentration. Both the Thomas and Bed Depth Service Time models successfully described the adsorption of Congo Red onto the jute fiber in a fixed bed column under different experimental conditions. After observing a satisfactory performance for the adsorption of Congo Red onto untreated jute (~ 2.8 mg/g, Table 2), the same research group further improved the adsorbent performance using chemical modifications (Roy, Adhikari and Majumder, 2013b). Namely, tannin-modified jute fibers were prepared in three steps: fiber surface activation with 1% NaOH, suspension of pretreated fibers in 0.25% NaOH in the presence of sequentially added epichlorohydrin, and immersing in 15 g/l tannin solution in the presence of $NaBH_4$ as a catalyst. The last two treatments last for a total of 8 h and are carried out at 60°C. Such fibers are capable of binding about 7.5 mg/g of Congo Red under the appropriate conditions, Table 2. It has to be mentioned that the adsorbent dose inversely affected the adsorption capacity of modified fibers, whereas the adsorption capacity increased with increasing initial dye concentration. The adsorption equilibrium data were found to follow the Langmuir isotherm model and the fast adsorption (15-30 min) conformed well with the pseudo-second-order kinetic model. The data obtained from the thermodynamic study prove that the adsorption of Congo Red onto tannin-modified jute fibers is of spontaneous and exothermic nature which is attributed to the myriad of available hydroxyl and other polar groups on the fiber surface. Based on the desorption experiments carried out in 0.1 M NaOH, the authors Roy, Adhikari and Majumder (2013b) concluded that the ion-exchange mechanism participates in the adsorption of Congo Red onto tannin-modified jute fibers. Almost a decade has passed since the first utilization of tannin-modified jute fibers as adsorbents for Congo Red from aqueous solution, and in 2021, Roy et al. (2022) used the same adsorbents for different types of azo, direct, reactive, disperse, and vat dyes present in real wastewater collected from the textile industry.

The previously discussed procedure for jute modification is long-lasting and complicated, while the adsorption potential of tannin-modified jute for Congo Red is somewhat lower than that recently reported (8.5 mg/g, Table 2) for jute fabric oxidized with 0.4% $NaIO_4$ for 120 min at room temperature (Ivanovska et al., 2022a). For comparison, untreated jute fabric has about 45% lower adsorption potential (Table 2) than oxidized one since the sodium periodate oxidation provides sufficient content of aldehyde groups and availability of naturally present carboxyl and hydroxyl groups, all of them participating in dye adsorption. The adsorption of Congo Red onto oxidized jute fabric followed the pseudo-second-order model indicating that the chemisorption process is primarily represented. On the other hand, the isotherm and thermodynamic studies pointed out that the adsorption is of spontaneous and endothermic character and obeys the Langmuir isotherm model. As is presented in Figure 5, a binding mechanism of Congo Red with oxidized jute proceeded *via* hydrogen bonds, repulsion, and π-π stacking interactions (Ivanovska et al., 2022a). The studied method for jute fabric modification is very promising having in mind a high adsorption potential of oxidized fabric as well as the possibility for fast regeneration of sodium periodate solution after the treatment (Koprivica et al. 2016).

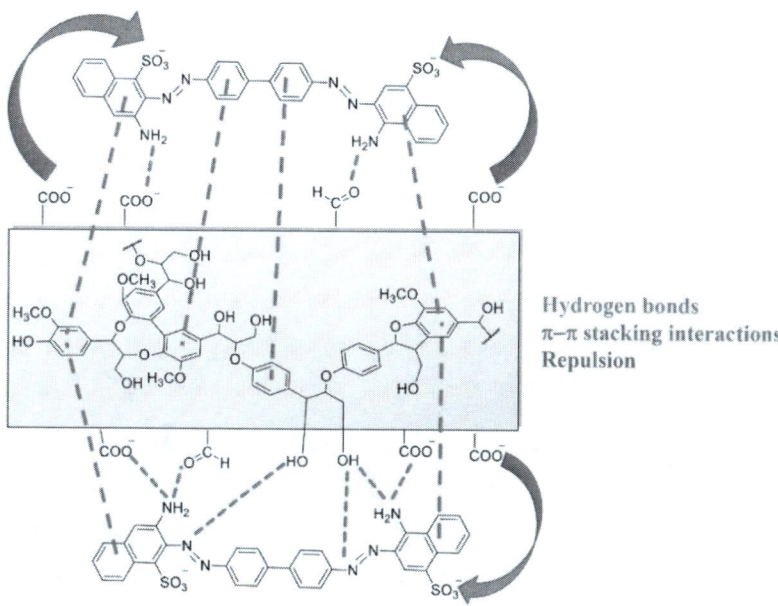

Figure 5. Proposed mechanism of Congo Red adsorption onto jute fabric oxidized with NaIO$_4$ (Ivanovska et al. 2022a). Copyright A. Ivanovska.

Among all studies in which Congo Red adsorption was investigated, the highest capacity of 33.7 mg/g was registered in the case of NaOH-modified jute fibers (Dey, Kumar and Dey, 2018) which could be ascribed to the significantly higher initial dye concentration as well as adsorbent dose, Table 2. Kinetic, isotherm, and thermodynamic experiments showed that the dye adsorption was entropy-driven and endothermic; it followed a pseudo-second-order model, and fit to Langmuir isotherm model. Almost the same group of authors (Dey and Dey, 2021) found that NaOH-treated jute fibers are valuable adsorbents for another azo dye, Reactive Red 195, whereby the experimentally obtained adsorption capacity was 32.2 mg/g, Table 2. The adsorption results depicted that the adsorbent dose along with the temperature have a great influence on the adsorption capacity.

The investigation conducted by Ivanovska et al. (2021c) is very specific since the alkali and oxidatively modified jute fabrics with enhanced sorption properties (i.e., water retention power, capillarity, and cross-sectional fiber swelling) were used as adsorbents for anthraquinone dye Acid Blue 111. The obtained maximum adsorption capacities for alkali and oxidatively modified jute fabrics are 16.4 and 17.0 mg/g, respectively, and the equilibrium adsorption data were highly consistent with the Langmuir isotherm model. Moreover, based on the predicted dye pKa values, the fabric zeta potential, the content of functional groups as well as hydrogen bond intensity, the authors proposed a possible mechanism of the dye adsorption onto jute fabric, Figure 6. More precisely, the fabric-dye interactions are established primarily through the hydrogen bonds between jute carboxyl groups and the dye N–H group, further strengthened by π-π stacking interactions between aromatic rings of both dye and jute fibers (Ivanovska et al., 2021c).

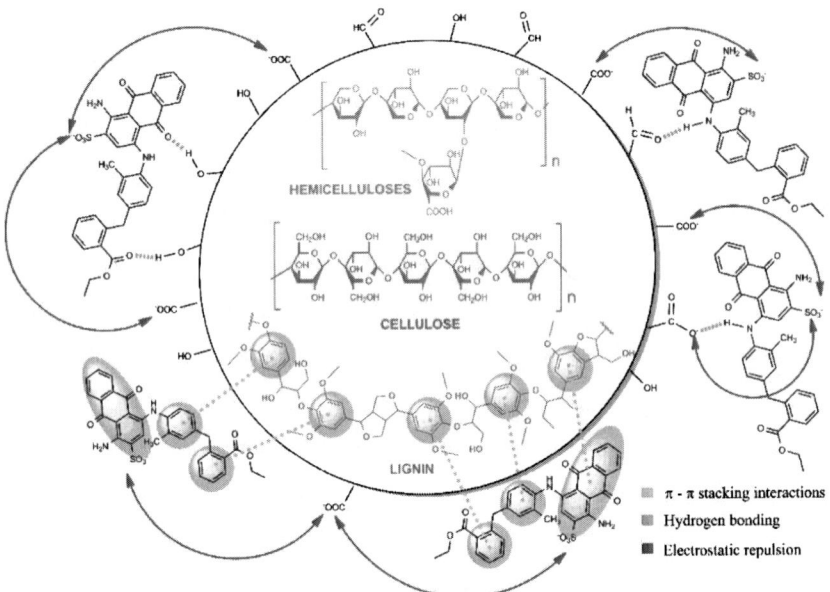

Figure 6. Proposed mechanism of Acid Blue 111 adsorption onto jute fabric (Ivanovska et al., 2021c) (Reprinted from Publication *Industrial Crops and Products*, 171, Ivanovska A, Lađarević J, Pavun L, Dojčinović B, Cvijetić I, Mijin D, Kostić M., Obtaining jute fabrics with enhanced sorption properties and "closing the loop" of their lifecycle, Article No. 113913, Copyright (2021), with permission from Elsevier).

High efficient jute adsorbent was prepared according to the following procedure: after the treatment with 5% NaOH for 2 h, washing, and drying, the fibers were treated with pyromellitic dianhydride (PMDA) in the solvent of N,N-dimethyl formamide (DMF) in the microwave reactor at 123°C for 25 min (Gao et al., 2015). Such modified fibers were washed, dried, and used as adsorbents for aniline, whereby the adsorption potential of 30.7 mg/g, was observed, Table 2. According to the presented mechanism, aniline was adsorbed onto modified jute fibers through hydrogen bonds as well as electrostatic interactions. Furthermore, based on the obtained results, the authors stated that the adsorption is spontaneous and endothermic, followed the pseudo-second-order kinetic model and Langmuir isotherm model. It has to be emphasized that the adsorbent could be regenerated through the desorption of aniline by using 0.5 M HCl solution, and the adsorption capacity after regeneration is even higher than that of virgin modified jute fibers.

After such good results, this research group continued to study the same adsorbent for aniline from an aqueous solution but in a fixed-bed column by varying bed depths, influent flow rates, and initial dye concentration (Hu et al., 2016). The adsorption capacity of 91.5 mg/g was obtained (initial dye concentration of 200 mg/l) at the bed depth of 24 mm. Dynamic modeling analysis revealed that the best predictions of breakthrough curves were obtained with the Bed Depth Service Time model. Again, the dye desorption was carried out in 0.5 M HCl and after three regeneration cycles, > 95% adsorption capacity of the jute fibers was maintained, which is very important for industrial-scale applications.

Besides different chemical modifications, other techniques were also used for improving the jute fibers' adsorption potential for synthetic dyes. For example, jute fibers were grafted with acrylic acid by gamma irradiation technique, and thereafter, chitosan was immobilized onto them (Hassan, 2015). Such modified fibers were used as adsorbents for Reactive Blue 222, Table 2. The performed experiments confirmed a surprisingly high adsorption potential of 220 mg/g. Moreover, modified jute fibers could be regenerated for up to three cycles, while after that, they lost about 50% of their dye adsorption efficiency (Hassan, 2015).

As in the case of heavy metal ions, the adsorption of synthetic dyes (such as Methylene Blue (Yousuf, Mahnaz and Syeda, 2021), Eosin Yellow, Malachite Green, and Crystal Violet (Porkodi and Vasanth Kumar, 2007), Acid Red 27 (Phan et al., 2006), Basic Blue 41 (Roy et al., 2022) and Reactive Red (Senthilkumaar et al., 2006) was also studied on activated carbons, whereby physically or chemically activated jute fibers were used a low-cost precursor. It should be underlined that the preparation of activated carbons usually implies treatments at high temperatures (400-1000°C) raising the question regarding the adsorbent cost-effectiveness.

According to the previously discussed results, it is clear that adsorption is a very suitable method for the decoloration of wastewaters and removal of synthetic dyes, primarily present in trace levels, i.e., below 100 mg/l, which covers the concentration of dyes found in the real textile effluents (10-60 mg/l) (Yaseen and Scholz, 2019). However, photocatalytic degradation is another technique that could be used for the same purpose. Thanks to the heterogeneity in chemical composition and structure, jute is a suitable substrate for the *in situ* synthesis of different nanoparticles having photocatalytic activity, which is intensively explored in the last five years. For example, Araújo et al. (2021) studied the photocatalytic activity of CaO and $CaO-SiO_2$ nanoparticles deposited on the jute fabrics, whereby Methylene Blue was selected as a model dye. Furthermore, jute fibers were explored as a biotemplate for the synthesis of ZnO nanoparticles with higher photocatalytic efficiency for the degradation of Crystal Violet dye under visible light (Abarna et al., 2016). Moreover, ZnO supported on porous biochar was synthesized through thermolysis of jute fibers and $Zn(OAc)_2$ in one pot and further used for dark adsorption and photocatalytic degradation of Methylene Blue (Chen et al., 2019). The highest efficiency of 99% was obtained at a pH of 7.0 following 30 min under UV illumination. Additionally, the ZnO nanorods/carbonized jute fiber composite was stable and showed an efficiency of over 80% during seven cycles of utilization and regeneration.

Possible Reuse and Revalorization of Saturated Jute Adsorbents

The major issue of the adsorption process is the adsorbent regeneration and the disposal of saturated adsorbents. After being applied as adsorbents for heavy metal ions or dyes, saturated jute materials are either regenerated for reuse or disposed of in landfills, which creates additional waste, i.e., secondary pollution. As it was discussed previously (Kumar, Chakraborty and Ray, 2008; Sahu, Mahapatra and Patel, 2017; Roy, Adhikari and Majumder, 2013b; Gao et al., 2015; Hu et al., 2016), the saturated adsorbent regeneration process requires the use of suitable agents for pollutant desorption like NaOH, or HCl. In such a way, new wastewaters are created, while the jute adsorption potential declines with an increase in the number of regeneration cycles. The last clearly explain why adsorbent regeneration is not applicable in real conditions.

Ivanovska et al. (2022a) suggested how the lifecycle of jute fabrics with adsorbed Congo Red could be extended. Namely, the saturated jute fabrics could be used as home textiles (i.e., decorative fabrics or carpet backing), or for the production of bio-based composites which would be used as building materials. Furthermore, saturated adsorbents can be carbonized and employed in various applications such as supercapacitors, Li-ion batteries, and electrocatalysts, owing to their high surface area with modulated pore size, low cost, and natural abundance (Veeraman et al., 2017). On the other hand, jute fabrics with adsorbed metal ions such as Ni^{2+} (Ivanovska et al., 2021b), Cu^{2+} and Cd^{2+} (De Quadros Melo et al., 2015), could be burned and the metals recovered (Mongioví et al., 2022b). The mentioned ways for extending the lifecycle of saturated adsorbents contributed to minimizing the waste accumulation and feedstock consumption and reducing the cost and environmental impacts of waste adsorbent disposal giving value to the waste (Huang et al., 2019).

To accomplish the concept "LESS WASTE, MORE VALUE," saturated jute adsorbents could be also used for obtaining some new materials which will have a new value. Having in mind that jute fibers have distinguished a hierarchical fibrilar composite microstructure and increasing efforts to convert these native biopolymeric materials into non-oxide as well as oxide ceramic products (Greil, 2001) or to obtain fiber-reinforced ceramics that combines the advantages of ceramics with an enhancement of mechanical properties (Alcaraz et al., 2019; Alzeer and MacKenzie, 2013) saturated jute sorbents can be used to produce such hybrid materials. The addition of fibers overcomes the ceramics materials' brittleness and increases their toughness and strength making such materials suitable for structural and construction applications. In such a way, adsorbed pollutants will remain immobilized in the structure of ceramic materials. Permanent collection and reuse of pollutant saturated jute materials have promising multi-positive effects on the economy as well environment, including reducing its quantity, saving energy, and its utilization as raw material for producing new hybrid materials which is in line with the Circular Economy Package (2020).

Conclusion

This chapter provides an overview of the possibility of the application of raw and chemically modified jute fibers as an eco-friendly adsorbent for heavy metals and dyes as the most frequent water pollutants. Special attention has been paid to the binding mechanisms of the pollutants and differently functionalized jute adsorbents. Due to the depletion of natural resources, increasing greenhouse emissions, and awareness of the need for sustainable development in terms of safely reusing waste, the transformation of saturated jute adsorbents into valuable materials is emerging as a strong trend. The last one represents one step toward both the circular economy approach and sustainable development, in terms of solid waste management, and the development of low-cost and eco-friendly wastewater treatment technologies. The presented results represent a significant step forward in the further development of advanced lignocellulosic fibers.

Acknowledgments

The authors would like to acknowledge financial support from the Ministry of Education, Science and Technological Development of the Republic of Serbia (grant numbers: 451-03-68/2022-14/200287, 451-03-68/2022-14/200135), and the Science Fund of the Republic of Serbia, #7726976, Integrated Strategy for Rehabilitation of Disturbed Land Surfaces and Control of Air Pollution - RECAP.

Disclaimer

None

References

Abarna B, Preethi T, Karunanithi A, Rajarajeswari GR. Influence of jute template on the surface, optical and photocatalytic properties of sol-gel derived mesoporous zinc oxide. *Materials Science in Semiconductor Processing* (2016) 56:243-250.

Alcaraz JS, Belda IM, Sanchis EJ, Gadea Borrell JM. Mechanical properties of plaster reinforced with yute fabrics. *Composites Part B: Engineering* (2019) 178:107390.

Al-Mamun M, Khan MA, Khan RA, Zaman HU, Saha M, Huque SMF. Preparation of selective ion adsorbent by photo curing with acrylic and phosphoric acid on jute yarn. *Fibers and Polymers*, (2010) 11:832-837.

Alzeer M, MacKenzie K. Synthesis and mechanical properties of novel composites of inorganic polymers (geopolymers) with unidirectional natural flax fibres (*phormium tenax*). *Applied Clay Science* (2013) 75-76:148-152.

Anderson A, Anbarasu A, Pasupuleti RR, Manigandan S, Praveenkumar TR, Aravind Kumar J. Treatment of heavy metals containing wastewater using biodegradable adsorbents: A review of mechanism and future trends. *Chemosphere* (2022) 295:133724.

Araújo JC, Ferreira DP, Teixeira P, Fangueiro R. In-situ synthesis of CaO and SiO_2 nanoparticles onto jute fabrics: exploring the multifunctionality. *Cellulose*, (2021) 28:1123-1138.

Baheti V, Padil VVT, Militký J, Černík M, Mishra R. Removal of mercury from aqueous environment by jute nanofiber. *Journal of Fiber Bioengineering and Informatics* (2013) 6:175-184.

Banerjee S, Dastidar MG. Use of jute processing wastes for treatment of wastewater contaminated with dye and other organics. *Bioresource Technology* (2005) 96:1919-1928.

Bjørklund G, Crisponi G, Nurchi VM, Buha Djordjevic A, Aaseth J. A review on coordination properties of thiol-containing chelating agents towards mercury, cadmium, and lead. *Molecules* (2019) 24:3247.

Buha A, Baralić K, Djukic-Cosic D, Bulat Z, Tinkov A, Panieri E, Saso L. The role of toxic metals and metalloids in Nrf2 signaling. *Antioxidants*, (2021) 10:630.

Calimli MH, Nas MS, Burhan H, Mustafov SD, Demirbas Ö, Sen F. Preparation, characterization and adsorption kinetics of methylene blue dye in reduced-graphene oxide supported nanoadsorbents. *Journal of Molecular Liquids* (2020) 309:113171.

Chakraborty R, Asthana A, Singh AK, Jain B, Susan, ABH. Adsorption of heavy metal ions by various low-cost adsorbents: a review. *International Journal of Environmental Analytical Chemistry* (2022) 102:342-379.

Chen M, Bao C, Hu D, Jin X, Huang Q. Facile and low-cost fabrication of ZnO/biochar nanocomposites from jute fibers for efficient and stable photodegradation of methylene blue dye. *Journal of Analytical and Applied Pyrolysis* (2019) 139:319-332.

Chen M, He F, Hu D, Bao C, Huang Q. Broadened operating pH range for adsorption/reduction of aqueous Cr(VI) using biochar from directly treated jute (*Corchorus capsularis* L.) fibers by H_3PO_4. *Chemical Engineering Journal*, (2020) 381:122739.

Circular Economy Package (2020) https://www.gov.uk/government/publications/circular-economy-package-policy-statement (Accessed August 5, 2022)

Crini, G. and Lichtfouse, E. (2019). Advantages and disadvantages of techniques used for wastewater treatment. *Environmental Chemistry Letters* 145-155.

Das KK, Reddy RC, Bagoji IB, Das S, Bagali S, Mullur L, Khodnapur JP, Biradar MS. Primary concept of nickel toxicity - an overview. *Journal of Basic and Clinical Physiology and Pharmacology* (2018) 30:141-152.

De Quadros Melo D, Vidal CB, Da Silva AL, Raulino GSC, Da Luz AD, Da Luz C, Fechine PBA, Mazzeto SE, Do Nascimento RF. Removal of toxic metal ions using modified lignocellulosic fibers as eco-friendly biosorbents: Mathematical modeling and numerical simulation. *International Journal of Civil and Environmental Engineering* (2015) 15:14-25.

Dey AK, Dey A. Selection of optimal processing condition during removal of Reactive Red 195 by NaOH treated jute fibre using adsorption. *Groundwater for Sustainable Development* (2021) 12:100522.

Dey AK, Kumar U, Dey A. Use of response surface methodology for the optimization of process parameters for the removal of Congo Red by NaOH treated jute fibre. *Desalination and Water Treatment* (2018) 115:300-314.

Du Z, Zheng T, Wang P, Hao L, Wang Y. Fast microwave-assisted preparation of a low-cost and recyclable carboxyl modified lignocellulose-biomass jute fiber for enhanced heavy metal removal from water. *Bioresource Technology* (2016) 201:41-49.

Du Z, Zheng T, Wang P. Experimental and modelling studies on fixed bed adsorption for Cu(II) removal from aqueous solution by carboxyl modified jute fiber. *Powder Technology* (2018) 338:952-959.

FAOSTAT - Food and Agriculture Organization of the United Nations (Accessed February 22, 2022)

Farooq MU, Jalees MI, Iqbal A, Zahra N, Kiran A. Characterization and adsorption study of biosorbents for the removal of basic cationic dye: kinetic and isotherm analysis. *Desalination and Water Treatment* (2019) 160:333-342.

Fodeke AA, Olayera OO. Thermodynamics of adsorption of malachite green hydrochloride on treated and untreated corncob charcoal. *Journal of the Serbian Chemical Society* (2019) 84:1143-1154.

Gao D-W, Hu Q, Pan H, Jiang J, Wang P. High-capacity adsorption of aniline using surface modification of lignocellulose-biomass jute fibers. *Bioresource Technology* (2015) 193:507-512.

Greil P. Biomorphous ceramics from lignocellulosics. *Journal of the European Ceramic Society*, (2001) 21:105-118.

Han G, Du Y, Huang Y, Wang W, Su S, Liu B. Study on the removal of hazardous Congo red from aqueous solutions by chelation flocculation and precipitation flotation process. *Chemosphere* (2022) 289:133109.

Hao L, Zheng T, Jiang J, Hu Q, Li X, Wang P. Removal of As(III) from water using modified jute fibres as a hybrid adsorbent. *RSC Advances* (2015) 5:10723-10732.

Hassan MS, Zohdy MH. Adsorption kinetics of toxic heavy metal ions from aqueous solutions onto grafted jute fibers with acrylic acid by gamma irradiation. *Journal of Natural Fibers* (2018) 15:506-516.

Hassan MS. Removal of reactive dyes from textile wastewater by immobilized chitosan upon grafted jute fibers with acrylic acid by gamma irradiation. *Radiation Physics and Chemistry* (2015) 115:55-61.

Hossain MdS, Rony FK, Sultana S, Kabir Md.H, Kabir SF, Ahmed S. Preparation and characterization of activated carbon from bagasse and jute fibre for heavy metal adsorption. *International Journal of Science and Management Studies* (2019) 2:17-34.

Hu Q, Wang P, Jiang J, Pan H, Gao D-W. Column adsorption of aniline by a surface modified jute fiber and its regeneration property. *Journal of Environmental Chemical Engineering* (2016) 4:2243-2249.

Huang Q, Hu D, Chen M, Bao C, Jin X. Sequential removal of aniline and heavy metal ions by jute fiber biosorbents: A practical design of modifying adsorbent with reactive adsorbate. *Journal of Molecular Liquids* (2019) 285:288-298.

Ivanovska A, Asanovic K, Jankoska M, Mihajlovski K, Pavun L, Kostic M. Multifunctional jute fabrics obtained by different chemical modifications. *Cellulose* (2020b) 27:8485-8502.

Ivanovska A, Branković I, Lađarević J, Pavun L, Kostic M. Oxidized jute as a valuable adsorbent for Congo Red from an aqueous solution. *Journal of Engineered Fibers and Fabrics* (2022a) 17:1-9.

Ivanovska A, Cerovic D, Maletic S, Jankovic Castvan I, Asanovic K, Kostic M. Influence of the alkali treatment on the sorption and dielectric properties of woven jute fabric. *Cellulose* (2019) 26:5133-5146.

Ivanovska A, Dojcinovic B, Maletic S, Pavun L, Asanovic K, Kostic M. Waste jute fabric as a biosorbent for heavy metal ions from aqueous solution. *Fibers and Polymers* (2020a) 21:1992-2002.

Ivanovska A, Lađarević J, Pavun L, Dojčinović B, Cvijetić I, Mijin D, Kostić M. Obtaining jute fabrics with enhanced sorption properties and "closing the loop" of their lifecycle. *Industrial Crops and Products*, (2021c) 171:113913.

Ivanovska A, Maletić S, Djokić V, Tadić N, Kostić M. Effect of chemical modifications and coating with Cu-based nanoparticles on the electro-physical properties of jute fabrics in a condition of high humidity. *Industrial Crops and Products* (2022b) 180:114792.

Ivanovska A, Pavun L, Dojčinović B, Kostić M. Kinetic and isotherm studies for the biosorption of nickel ions by jute fabrics. *Journal of the Serbian Chemical Society* (2021b) 86:885-897.

Ivanovska A, Veljović S, Dojčinović B, Tadić N, Mihajlovski K, Natić M, Kostić M. A strategy to revalue a wood waste for simultaneous cadmium removal and wastewater disinfection. *Adsorption Science and Technology*, (2021a) 2021:3552300.

Khera RA, Iqbal M, Jabeen S, Abbas M, Nazir A, Nisar J, Ghaffar A, Shar GA, Tahir MA. Adsorption efficiency of Pitpapra biomass under single and binary metal systems. *Surfaces and Interfaces* (2019) 14:138-145.

Koprivica S, Siller M, Hosoya, T, Roggenstein W, Rosenau T, Potthast A.Regeneration of aqueous periodate solutions by ozone treatment: A sustainable approach for dialdehyde cellulose production. *ChemSusChem*, (2016) 9:825-833.

Kukić D, Ivanovska A, Vasić V, Lađarević J, Kostić M, Šćiban M. The overlooked potential of raspberry canes: from waste to an efficient low-cost biosorbent for Cr(VI) ions. *Biomass Conversion and Biorefinery*, (2022) in press. https://doi.org/10.1007/s13399-022-02502-4.

Kumar PA, Chakraborty S, Ray M. Removal and recovery of chromium from wastewater using short chain polyaniline synthesized on jute fiber. *Chemical Engineering Journal* (2008) 141:130-140.

Kumar PA, Chakraborty S. Fixed-bed column study for hexavalent chromium removal and recovery by short-chain polyaniline synthesized on jute fiber. *Journal of Hazardous Materials* (2009) 162:1086-1098.

Kumar V, Parihar RD, Sharma A, Bakshi P, Singh Sidhu GP, Bali AS, Karaouzas I, Bhardwaj R, Thukral AK, Gyasi-Agyei Y, Rodrigo-Comino J. Global evaluation of heavy metal content in surface water bodies: A meta-analysis using heavy metal pollution indices and multivariate statistical analyses. *Chemosphere* (2019) 236:124364.

Laizer AGK, Bidu JM, Selmani JR, Njau KN. Improving biological treatment of textile wastewater. *Water Practice and Technology* (2022) 17:456-468.

Loiacono S, Crini G, Chanet G, Raschetti M, Placet V, Morin-Crini N. Metals in aqueous solutions and real effluents: biosorption behavior of a hemp-based felt. *Journal of Chemical Technology and Biotechnology*, (2018) 93:2592-2601.

Meng Z, Bai X, Tang X. Short-term assessment of heavy metals in surface water from Xiaohe river irrigation area, China: Levels, sources and distribution. *Water*, (2022) 14:1273.

Mladenovic N, Makreski P, Tarbuk A, Grgic K, Boev B, Mirakovski D, Toshikj E, Dimova V, Dimitrovski D, Jordanov I. Improved dye removal ability of modified rice husk with effluent from alkaline scouring based on the circular economy concept. *Processes* (2020) 8:653.

Mongioví C, Lacalamita D, Morin-Crini N, Gabrion X, Placet V, Lado Ribeiro AR, Ivanovska A, Kostić M, Bradu C, Staelensg J-N, Martel B, Raschetti M, Crini G. Use of chènevotte, a valuable co-product of industrial hemp fiber, as adsorbent for copper ions: Kinetic studies and modeling. *Arabian Journal of Chemistry* (2022) 15:103742.

Mongioví C, Morin-Crini N, Placet V, Bradu C, Ribeiro ARL, Ivanovska A, Kostić M, Martel B, Cosentino C, Torri G, Rizzi V, Gubitosa J, Fini P, Cosma P, Lichtfouse E, Lacalamita D, Mesto E, Schingaro E, De Vietro N, Crini G. Hemp-based materials for applications in wastewater treatment by biosorption-oriented processes: A review. In: *Cannabis/Hemp for Sustainable Agriculture and Materials* (2022b) 239-295.

Nurchi VM, Buha Djordjevic A, Crisponi G, Alexander J, Bjørklund G, Aaseth J. Arsenic toxicity: Molecular targets and therapeutic agents. *Biomolecules* (2020) 10:235.

Perumal M, Karikalacholan S, Parimannan N, Arichandran J, Shanmuganathan K, Ravi R, Jayapandiyan S, Jayakumar S, Mohandas T. Chapter 20 - Integrated anaerobic-aerobic processes for treatment of high strength wastewater: Consolidated application, new trends, perspectives, and challenges. *Integrated Environmental Technologies for Wastewater Treatment and Sustainable Development* (2022) 457-481.

Phan NH, Rio S, Faur C, Le Coq L, Le Cloirec P, Nguyen TH. Production of fibrous activated carbons from natural cellulose (jute, coconut) fibers for water treatment applications. *Carbon* (2006) 44:2569-2577.

Porkodi K, Vasanth Kumar K. Equilibrium, kinetics and mechanism modeling and simulation of basic and acid dyes sorption onto jute fiber carbon: Eosin yellow, malachite green and crystal violet single component systems. *Journal of Hazardous Materials* (2007) 143:311-327.

Prajapati K, Sidhpuria K, Mahajan D, Chakraborty M. Studies on equilibrium and kinetics of ACRY Red 4G removal from aqueous solutions using low cost adsorbents. *Indian Journal of Chemical Technology* (2005) 12:425-429.

Ramim SS, Sultana H, Akter T, Ali MA. Removal of arsenic from groundwater using iron-coated jute-mesh structure. *Desalination and Water Treatment* (2017) 100:347-353.

Roy A, Adhikari B, Majumder SB. Equilibrium, kinetic, and thermodynamic studies of azo dye adsorption from aqueous solution by chemically modified lignocellulosic jute fiber. *Industrial and Engineering Chemistry Research* (2013b) 52:6502-6512.

Roy A, Chakraborty S, Kundu SP, Adhikari B, Majumder SB. Lignocellulosic jute fiber as a bioadsorbent for the removal of azo dye from its aqueous solution: Batch and column studies. *Journal of Applied Polymer Science* (2013a) 129:15-27.

Roy A. Removal of color from real textile dyeing effluent utilizing tannin immobilized jute fiber as biosorbent: optimization with response surface methodology. *Environmental Science and Pollution Research* (2021) 28:12011-12025.

Roy H, Shakil R, Tarek YA, Firoz SH. Study of the removal of Basic blue-41 from simulated wastewater by activated carbon prepared from discarded jute fiber. *ECS Transactions* (2022) 107:8407-9420.

Sahu UK, Mahapatra SS, Patel RK. Synthesis and characterization of an eco-friendly composite of jute fiber and Fe_2O_3 nanoparticles and its application as an adsorbent for removal of As(V) from water. *Journal of Molecular Liquids* (2017) 237:313-321.

Sarkar PB, Chatterjee H, Mazumdar AK. Absorption of basic dyes by jute. *Nature* (1946) 157:486.

Sarkar PB, Chatterjee H. Studies on the absorption of methylene blue by jute fibre. *Journal of the Society of Dyers and Colourists* (1948) 218-221.

Senthilkumaar S, Kalaamani P, Porkodi K, Varadarajan PR, Subburaam CV. Adsorption of dissolved Reactive red dye from aqueous phase onto activated carbon prepared from agricultural waste. *Bioresource Technology* (2006) 97:1618-1625.

Shukla SR, Pai RS. Adsorption of Cu(II), Ni(II) and Zn(II) on modified jute fibers. *Bioresource Technology* (2005a) 96:1430-1438.

Shukla SR, Pai RS. Removal of Pb(II) from solution using cellulose-containing materials. *Journal of Chemical Technology and Biotechnology* (2005b) 80:176-183.

Shukla SR, Sakhardande VD. Column studies on metal ion removal by dyed cellulosic materials. *Journal of Applied Polymer Science* (1992) 44:903-910.

Shukla SR, Sakhardande VD. Metal ion removal by dyes cellulosic materials. *Journal of Applied Polymer Science* (1991a) 42:829-835.

Shukla SR, Sakhardande VD. Removal of metal ions using dyed cellulosic materials. *Dyes and Pigments* (1991b) 17:101-112.

Veeramani V, Sivakumar M, Chen S-M, Madhu R, Alamri HR, Alothman ZA, Hossain MSA, Chen C-K, Yamauchi Y, Miyamoto N, Wu KC-W. Lignocellulosic biomass-derived, graphene sheet-like porous activated carbon for electrochemical supercapacitor and catechin sensing. *RSC Advances* (2017) 72:45668-45675.

Yaseen DA, Scholz M. Textile dye wastewater characteristics and constituents of synthetic effluents: a critical review. *International Journal of Environmental Science and Technology* (2019) 16:1193-1226.

Yousuf MdR, Mahnaz F, Syeda SR. Activated carbon fiber from natural precursors: a review of preparation methods with experimental study on jute fiber. *Desalination and Water Treatment* (2021) 213:441-458.

Zhou Q, Yang N, Li Y, Ren B, Ding X, Bian H, Yao X. Total concentrations and sources of heavy metal pollution in global river and lake water bodies from 1972 to 2017. *Global Ecology and Conservation* (2020) 22:e00925.

Biographical Sketches

Aleksandra Ivanovska

Affiliation: Innovation Center of the Faculty of Technology and Metallurgy in Belgrade Ltd., University of Belgrade

Education:
2007-2011: BSc. Textile Engineering, Ss. Cyril and Methodius University in Skopje, Faculty of Technology and Metallurgy
2012-2014: MSc. Chemical Textile Technology and Ecology, Ss. Cyril and Methodius University in Skopje, Faculty of Technology and Metallurgy
2015-2022: PhD in Science, University of Belgrade, Faculty of Technology and Metallurgy

Business Address: Karnegijeva 4, 11000 Belgrade, Serbia

Research and Professional Experience:
In October 2018 (1 month), she has been a visiting researcher in the Textile laboratory at the University Ss. Cyril and Methodius University in Skopje, Faculty of Technology and Metallurgy

Project member:
2017-2019: Functionalization, characterization and application of cellulose and cellulose derivatives (ON172019), funded by the Ministry of Education, Science and Technological Development of the Republic of Serbia (Project Leader Mirjana Kostić)
2018-2019: Functional cellulose based clothing promoting healthier well-being wear comfort for immobile people (Ev. No: 47), Serbian-Slovenian Bilateral Project (Project Leaders Ana Kramar and Lidija Fras Zemljič)

2019-2022: Sustainable coloration process of protective fabrics based on novel dye architecture with distinctive properties (Project ID 5540), funded by the Innovation Fund of the Republic of Serbia (Project Leader Jelena Lađarević)
2021-now: Compatibility in dispersion of surface biomaterial - Treated filler with biosourced polymers, funded by OMYA International AG Switzerland (Project Leaders Đorđe Janaćković and Petar Uskoković)
2022-now: Integrated Strategy for Rehabilitation of Disturbed Land Surfaces and Control of Air Pollution, funded by Science Fund of the Republic of Serbia (Project Leader Zorica Svirčev)

Professional Appointments:
2017-2019: Junior Research Assistant, Innovation Center of the Faculty of Technology and Metallurgy in Belgrade Ltd., University of Belgrade
2019-2020: Research Assistant, Innovation Center of the Faculty of Technology and Metallurgy in Belgrade Ltd., University of Belgrade
2020-now: Research Associate, Innovation Center of the Faculty of Technology and Metallurgy in Belgrade Ltd., University of Belgrade

Publications from the Last 3 Years:
1. Ivanovska A., Cerovic D., Maletic S., Jankovic Castvan I., Asanovic K., Kostic M.: Influence of the alkali treatment on the sorption and dielectric properties of woven jute fabric. *Cellulose,* Vol. 26, No. 8, 2019, pp. 5133-5146. https://doi.org/10.1007/s10570-019-02421-0.
2. Ivanovska A., Cerovic D., Tadic N., Jankovic Castvan I., Asanovic K., Kostic M.: Sorption and dielectric properties of jute woven fabrics: Effect of chemical composition. *Industrial Crops and Products,* Vol. 140, 2019, Article number: 111632. https://doi.org/10.1016/j.indcrop.2019.111632.
3. Ivanovska A., Asanovic K., Jankoska M., Mihajlovski K., Pavun L., Kostic M.: Multifunctional jute fabrics obtained by different chemical modifications. *Cellulose,* Vol. 27, No. 8, 2020, pp. 8485-8502. https://doi.org/10.1007/s10570-020-03360-x.
4. Ivanovska A., Dojcinovic B., Maletic S., Pavun L., Asanovic K., Kostic M.: Waste jute fabric as a biosorbent for heavy metal ions from aqueous

solution, *Fibers and Polymers*, Vol. 21, No. 9, 2020, pp. 1992-2020. https://doi.org/10.1007/s12221-020-9639-8.
5. Asanovic K. A., Cerovic D. D., Kostic M. M., Mihailovic T. V., Ivanovska A. M.: Multipurpose nonwoven viscose/polypropylene fabrics: Effect of fabric characteristics and humidity conditions on the volume electrical resistivity and dielectric loss tangent, *Fibers and Polymers*, Vol. 21, No. 10, 2020, pp. 2407-2416. https://doi.org/10.1007/s12221-020-1340-4.
6. Ivanovska A., Kostić M.: Electrokinetic properties of chemically modified jute fabrics. *Journal of the Serbian Chemical Society*, Vol. 85, No. 12, 2020, pp. 1621-1627. https://doi.org/10.2298/JSC201013069I.
7. Kramar A., Ivanovska A., Kostic M.: Regenerated cellulose fiber functionalization by two-step oxidation using sodium periodate and sodium chlorite - Impact on the structure and sorption properties. *Fibers and Polymers*, Vol. 22, No. 8, 2021. https://doi.org/10.1007/s12221-021-0996-8.
8. Ivanovska A., Reljic M., Kostic M., Asanovic K., Mangovska B.: Air permeability and water vapor resistance of differently finished cotton and cotton/elastane single jersey knitted fabrics. *Journal of Natural Fibers* (Article in Press). https://doi.org/10.1080/15440478.2021.1875383
9. Asanovic K., Cerovic D., Kostic M., Maletic S., Ivanovska A.: Electrophysical properties of woven clothing fabrics before and after abrasion. *Journal of Natural Fibers* (Article in Press). https://doi.org/10.1080/15440478.2021.1921659.
10. Ivanovska A., Lađarević J., Pavun L., Dojčinović B., Cvijetić I., Mijin D., Kostić M.: Obtaining jute fabrics with enhanced sorption properties and "closing the loop" of their lifecycle. *Industrial Crops and Products*, Vol. 171, 2021, Article number: 113913. https://doi.org/10.1016/j.indcrop.2021.113913.
11. Mašulović A. D., Lađarević J. M., Ivanovska A. M., Stupar S. Lj., Vukčević M. B., Kostić M. M., Mijin D. Ž.: Structural insight into the fiber dyeing ability: Pyridinium arylazo pyridone dyes. *Dyes and Pigments*, Vol. 195, Article number: 109741, 2021. https://doi.org/10.1016/j.dyepig.2021.109741.
12. Ivanovska A., Pavun L., Dojčinović B., Kostić M.: Kinetic and isotherm studies for the biosorption of nickel ions by jute fabrics. *Journal of the Serbian Chemical Society*, Vol. 86, No. 9, pp. 885-897, 2021. https://doi.org/10.2298/JSC210209030I.

13. A strategy to revalue a wood waste for simultaneous cadmium removal and wastewater disinfection. *Adsorption Science and Technology,* Vol. 2021, Article ID: 3552300, 2021. https://doi.org/10.1155/2021/3552300.
14. Mongioví C., Morin-Crini N., Lacalamita D., Bradu C., Raschetti M., Placet V., Ribeiro A.R.L., Ivanovska A., Kostić M., Crini G.: Biosorbents from plant fibers of hemp and flax for metal removal: Comparison of their biosorption properties. *Molecules,* Vol. 26, No. 14, 2021, Article No. 4199. https://doi.org/10.3390/molecules26144199.
15. Mongioví C., Lacalamita D., Morin-Crini N., Gabrion X., Ivanovska A., Sala F., Placet V., Rizzi V., Gubitosa J., Mesto E., Ribeiro A.R.L., Fini P., De Vietro N., Schingaro E., Kostić M., Cosentino C., Cosma P., Bradu C., Chanet G., Crini G.: Use of chènevotte, a valuable co-product of industrial hemp fiber, as adsorbent for pollutant removal. Part I: Chemical, microscopic, spectroscopic and thermogravimetric characterization of raw and modified samples. *Molecules,* Vol. 26, No. 14, 2021, Article No. 4199. https://doi.org/10.3390/molecules26144199.
16. Ivanovska A., Veljović S., Reljić M., Lađarević J., Pavun L., Natić M, Kostić M: Closing the loop: Dyeing and adsorption potential of mulberry wood waste, *Journal of Natural Fibers,* 2021 (Article in Press). https://doi.org/10.1080/15440478.2021.2009398.
17. Asanovic A. K., Ivanovska M. A., Jankoska Z. M., Bukhonka N., Mihailovic T., Kostic M. M.: Influence of pilling on the quality of flax single jersey knitted fabrics. *Journal of Engineered Fibers and Fabrics.* Vol. 17, 2022, pp. 1-13. https://doi.org/10.1177/15589250221091267.
18. Ivanovska A., Branković I., Lađarević J., Pavun L., Kostic M.: Oxidized jute as a valuable adsorbent for Congo Red from an aqueous solution. *Journal of Engineered Fibers and Fabrics.* Vol. 17, 2022, pp. 1-9. https://doi.org/10.1177/15589250221101380.
19. Mongioví C., Lacalamita D., Morin-Crini N., Gabrion X., Placet V., Lado Ribeiro A. R., Ivanovska A., Kostić M., Bradu C., Staelensg J-N., Martel B, Raschetti M., Crini G.: Use of chènevotte, a valuable co-product of industrial hemp fiber, as adsorbent for copper ions: Kinetic studies and modeling. *Arabian Journal of Chemistry,* Vol. 15, No. 4, 2022, Article ID 103742, https://doi.org/10.1016/j.arabjc.2022.103742.
20. C. Mongioví, N. Morin-Crini, V. Placet, C. Bradu, ARL. Ribeiro, A. Ivanovska, M. Kostić, B. Martel, C. Cosentino, G. Torri, V. Rizzi, J. Gubitosa, P. Fini, P. Cosma, E. Lichtfouse, D. Lacalamita, E. Mesto, E. Schingaro, N. De Vietro, G. Crini. Hemp-based materials for applications in wastewater treatment by biosorption-oriented processes: A review. In:

Cannabis/Hemp for Sustainable Agriculture and Materials. D.C. Agrawal, R. Kumar and M. Dhanasekaran, eds. Springer Nature Singapore Pte Ltd., chapter 9, pp. 239-295, 2022. https://doi.org/10.1007/978-981-16-8778-5_9.

21. Ivanovska A., Maletić S., Djokić V., Tadić N., Kostić M.: Effect of chemical modifications and coating with Cu-based nanoparticles on the electro-physical properties of jute fabrics in a condition of high humidity. *Industrial Crops and Products,* Vol. 180, 2022, Article number: 114792. https://doi.org/10.1016/j.indcrop.2022.114792.

22. Ivanovska A., Ladarević J., Asanović K., Barać N., Mihajlovski K., Kostić M., Mangovska B.: Quality of cotton and cotton/elastane single jersey knitted fabrics before and after softening and in situ synthesis of Cu-based nanoparticles. *Journal of Natural Fibers* (Article in Press) 2022. https://doi.org/10.1080/15440478.2022.2070328.

23. Kostic M., Imani M., Ivanovska A., Radojevic V., Dimic-Misic K., Barac N., Stojanovic D., Janackovic Dj., Uskokovic P., Barcelo E., Gane P.: Extending waste paper, cellulose and filler use beyond recycling by entering the circular economy creating cellulose-$CaCO_3$ composites reconstituted from ionic liquid. *Cellulose,* Vol. 29, No. 8, 2022, pp. 5037-5059. https://doi.org/10.1007/s10570-022-04575-w.

24. Kukić D., Ivanovska A., Vasić V., Ladarević J., Kostić M., Šćiban M.: The overlooked potential of raspberry canes: from waste to an efficient low-cost biosorbent for Cr(VI) ions. *Biomass Conversion and Biorefinery,* 2022, (Article in Press) https://doi.org/10.1007/s13399-022-02502-4.

25. Imani M., Dimic-Misic K., Kostic M., Barac N., Janackovic D., Uskokovic P., Ivanovska A., Lahti J., Barcelo E., Gane P.: Achieving a superhydrophobic, moisture, oil and gas barrier film using a regenerated cellulose-calcium carbonate composite derived from paper components or waste. *Sustainability,* Vol. 14, No. 16, 2022, Article number: 10425 2022. https://doi.org/10.3390/su141610425.

26. Ivanovska A., Asanović K., Jankoska M., Pavlović S., Poparić G., Kostić M.: Alkali treated jute fabrics suitable for the production of inexpensive technical textiles. Accepted for Publication in *Fibers and Polymers*.

27. Ivanovska A., Ladarević J., Asanović K., Pavun L., Kostić M., Mangovska B.: Revalorization of cotton and cotton/elastane knitted fabric waste. Accepted for publication in *Fibers and Polymers*.

Mirjana Kostic

Affiliation: Faculty of Technology and Metallurgy, University of Belgrade

Education:
1994-1998: PhD Chemistry and Chemical Technology, University of Belgrade, Faculty of Technology and Metallurgy
1989-1993: MSc. Technical Science, University of Belgrade, Faculty of Technology and Metallurgy
1984-1989: BSc. Chemical Technology, University of Belgrade, Faculty of Technology and Metallurgy

Business Address: Karnegijeva 4, 11000 Belgrade, Serbia

Research and Professional Experience:
02.-09.2010: Visiting researcher, Department of Chemistry, University of Natural Resources and Life Science, Vienna, Austria
09.2004-08.2005: Visiting researcher, Institute of Chemistry, University of Natural Resources and Life Sciences, Vienna, Austria
04.-10.2003: Visiting researcher, Institute of Chemistry, University of Natural Resources and Life Sciences, Vienna, Austria

Project member:
2014-2015: Permanent polysaccharide binding onto oxidised cellulose fibres providing antimicrobial effect (Ev.No:451-03-3095/2014-09/25), Serbian-Slovenian bilateral project (Project Leaders M. Kostic and L. Fras Zemljič).
2011-2019: Functionalization, characterization and application of cellulose and cellulose derivatives (ON172019), funded by the Ministry of Education, Science and Technological Development of the Republic of Serbia (Project Leader)
2018-2019: Functional cellulose based clothing promoting healthier well-being wear comfort for immobile people (Ev. No: 47), Serbian-Slovenian bilateral project (Project Leaders Ana Kramar and L. Fras Zemljič)
2019-2022: Sustainable coloration process of protective fabrics based on novel dye architecture with distinctive properties (Project ID 5540), funded by the Innovation Fund of the Republic of Serbia (Project Leader Jelena Ladarević)

2021-now: Compatibility in dispersion of surface biomaterial - Treated filler with biosourced polymers, funded by OMYA International AG Switzerland (Project Leaders Đorđe Janaćković and Petar Uskoković)
2022-now: Integrated Strategy for Rehabilitation of Disturbed Land Surfaces and Control of Air Pollution, funded by Science Fund of the Republic of Serbia (Project Leader Zorica Svirčev)

Professional Appointments:
2021-now: Vice-Dean, Faculty of Technology and Metallurgy, University of Belgrade, Serbia.
2014-now: Full Professor, Department of Textile Engineering, Faculty of Technology and Metallurgy, University of Belgrade, Serbia.
2011-2021: Head of Department of Textile Engineering, Faculty of Technology and Metallurgy, University of Belgrade
2009-2014: Associate Professor, Department of Textile Engineering, Faculty of Technology and Metallurgy, University of Belgrade, Serbia.
1999-2009: Assistant Professor, Department of Textile Engineering, Faculty of Technology and Metallurgy, University of Belgrade, Serbia.
1990-1999: Teaching Assistant, Department of Textile Engineering, Faculty of Technology and Metallurgy, University of Belgrade, Serbia

Publications from the Last 3 Years:
1. Ivanovska A., Cerovic D., Maletic S., Jankovic Castvan I., Asanovic K., Kostic M.: Influence of the alkali treatment on the sorption and dielectric properties of woven jute fabric. *Cellulose*, Vol. 26, No. 8, 2019, pp. 5133-5146. https://doi.org/10.1007/s10570-019-02421-0.
2. Ivanovska A., Cerovic D., Tadic N., Jankovic Castvan I., Asanovic K., Kostic M.: Sorption and dielectric properties of jute woven fabrics: Effect of chemical composition. *Industrial Crops and Products*, Vol. 140, 2019, Article number: 111632. https://doi.org/10.1016/j.indcrop.2019.111632.
3. Dimic-Misic K., Kostić M., Obradović B., Kramar A., Jovanović S., Stepanenko D., Mitrović-Dankulov M., Lazović S., Johansson L.-S., Maloney T., Gane P., Nitrogen plasma surface treatment for improving polar ink adhesion on micro/nanofibrillated cellulose films. *Cellulose*, 26, 2019, pp. 3845-3857. https://doi.org/10.1007/s10570-019-02269-4.
4. Korica M., Fras Zemljič L., Bračič M., Kargl R., Spirk S., Reishofer D., Mihajlovski K., Kostić M., Novel protein-repellent and antimicrobial

polysaccharide multilayer thin films, *Holzforschung.* 73(1), 2019, pp. 93-103. https://doi.org/10.1515/hf-2018-0094.
5. Morin-Crini N., Loiacono S., Placet V., Torri G., Bradu C., Kostić M., Cosentino C., Chanet G., Martel B., Lichtfouse E., Crini G., Hemp-based adsorbents for sequestration of metals: a review. *Environmental Chemistry Letters,* 17(1), 2019, pp. 393-408. https://doi.org/10.1007/s10311-018-0812-x.
6. Korica M., Peršin Z., Trifunović S., Mihajlovski K., Nikolić T., Maletić S., Fras Zemljič L., Kostić M.M., Influence of different pretreatments on the antibacterial properties of chitosan functionalized viscose fabric: TEMPO oxidation and coating with TEMPO oxidized cellulose nanofibrils. *Materials,* 12(19), 2019, pp. 3144. https://doi.org/10.3390/ma12193144.
7. Asanovic K.A., Kostic M.M., Mihailovic T.V., Cerovic D.D., Compression and strength behaviour of viscose/polypropylene nonwoven fabrics. *Indian Journal of Fibre & Textile Research,* 44 (3), 2019, pp. 329-337.
8. Ivanovska A., Asanovic K., Jankoska M., Mihajlovski K., Pavun L., Kostic M.: Multifunctional jute fabrics obtained by different chemical modifications. *Cellulose,* Vol. 27, No. 8, 2020, pp. 8485-8502. https://doi.org/10.1007/s10570-020-03360-x.
9. Ivanovska A., Dojcinovic B., Maletic S., Pavun L., Asanovic K., Kostic M.: Waste jute fabric as a biosorbent for heavy metal ions from aqueous solution, *Fibers and Polymers,* Vol. 21, No. 9, 2020, pp. 1992-2020. https://doi.org/10.1007/s12221-020-9639-8.
10. Asanovic K.A., Cerovic D.D., Kostic M.M., Mihailovic T.V., Ivanovska A.M.: Multipurpose nonwoven viscose/polypropylene fabrics: Effect of fabric characteristics and humidity conditions on the volume electrical resistivity and dielectric loss tangent, *Fibers and Polymers,* Vol. 21, No. 10, 2020, pp. 2407-2416. https://doi.org/10.1007/s12221-020-1340-4.
11. Ivanovska A., Kostić M.: Electrokinetic properties of chemically modified jute fabrics. *Journal of the Serbian Chemical Society,* Vol. 85, No. 12, 2020, pp. 1621-1627. https://doi.org/10.2298/JSC201013069I.
12. Milanovic J., Schiehser S., Potthast A., Kostic M., Stability of TEMPO-oxidized cotton fibers during natural aging. *Carbohydrate Polymers,* 230, 2020, pp. 115587. https://doi.org/10.1016/j.carbpol.2019.115587.
13. Pejić B.M., Kramar A.D., Obradović B.M., Kuraica M.M., Žekić A.A., Kostić M.M., Effect of plasma treatment on chemical composition, structure and sorption properties of lignocellulosic hemp fibers (*Cannabis*

sativa L.). *Carbohydrate Polymers*, 236, 2020, pp. 116000. https://doi.org/10.1016/j.carbpol.2020.116000.
14. Milanovic J., Lazic T., Zivkovic I., Vuksanovic M., Milosevic M., Kostic M., The effect of nanofibrillated tempo-oxidized cotton linters on the strength and optical properties of paper. *Journal of Natural Fibers*, 2020 (Article in Press). https://doi.org/10.1080/15440478.2020.1848742.
15. Sulaeva I., Hettegger H., Bergen A., Rohrer C., Kostic M., Konnerth J., Rosenau T., Potthast A., Fabrication of bacterial cellulose-based wound dressings with improved performance by impregnation with alginate. *Materials Science & Engineering C* 110, 2020, pp. 110619, https://doi.org/10.1016/j.msec.2019.110619.
16. Milanovic P.M., Stankovic S.B., Novakovic M., Grujic D., Kostic M., Milanovic J.Z., Development of the automated software and device for determination of wicking in textiles using open-source tools. *PLoS ONE*, 15(11), 2020, Article No.: e0241665, https://doi.org/10.1371/journal.pone.0241665.
17. Kodrić M., Đorđević D., Konstantinović S., Kostić M., Šarac T., Modeling of disperse dye adsorption on modified polyester fibers. *Acta Periodica Technologica*, Vo. 51, 2020, pp. 1-7
18. Kramar A., Ivanovska A., Kostic M.: Regenerated cellulose fiber functionalization by two-step oxidation using sodium periodate and sodium chlorite - Impact on the structure and sorption properties. *Fibers and Polymers*, Vol. 22, No. 8, 2021. https://doi.org/10.1007/s12221-021-0996-8.
19. Ivanovska A., Reljic M., Kostic M., Asanovic K., Mangovska B.: Air permeability and water vapor resistance of differently finished cotton and cotton/elastane single jersey knitted fabrics. *Journal of Natural Fibers* (Article in Press). https://doi.org/10.1080/15440478.2021.1875383.
20. Asanovic K., Cerovic D., Kostic M., Maletic S., Ivanovska A.: Electro-physical properties of woven clothing fabrics before and after abrasion. *Journal of Natural Fibers* (Article in Press). https://doi.org/10.1080/15440478.2021.1921659.
21. Ivanovska A., Lađarević J., Pavun L., Dojčinović B., Cvijetić I., Mijin D., Kostić M.: Obtaining jute fabrics with enhanced sorption properties and "closing the loop" of their lifecycle. *Industrial Crops and Products*, Vol. 171, 2021, Article number: 113913. https://doi.org/10.1016/j.indcrop.2021.113913.
22. Mašulović A.D., Lađarević J.M., Ivanovska A.M., Stupar S. Lj., Vukčević M.B., Kostić M.M., Mijin D.Ž.: Structural insight into the fiber dyeing

ability: Pyridinium arylazo pyridone dyes. *Dyes and Pigments*, Vol. 195, Article number: 109741, 2021. https://doi.org/10.1016/j.dyepig.2021.109741.
23. Kramar, A.D., Ilic-Tomic, T.R., Ladarević, J.M., Nikodinovic-Runic, J.B., Kostic, M.M., Halochromic cellulose textile obtained via dyeing with biocolorant isolated from *Streptomyces sp.* strain NP4. *Cellulose*, 28(13), 2021, pp. 8771-8784. https://doi.org/10.1007/s10570-021-04071-7.
24. Lazic B.D., Janjic S.D., Korica M., Pejic B.M., Djokic V.R., Kostic M.M., Electrokinetic and sorption properties of hydrogen peroxide treated flax fibers (*Linum usitatissimum* L.). *Cellulose*, 28, 2021, pp. 2889-2903. https://doi.org/10.1007/s10570-021-03686-0.
25. Kramar A.D., Obradović B.M., Schiehser S., Potthast A., Kuraica M.M., Kostić M.M., Enhanced antimicrobial activity of atmospheric pressure plasma treated and aged cotton fibers. *Journal of Natural Fibers*, 2021 (Article in Press). https://doi.org/10.1080/15440478.2021.1946883.
26. Milanovic J.Z., Milosevic M., Jankovic-Castvan I., Kostic M.M., Capillary rise and sorption ability of hemp fibers oxidized by non-selective oxidative agents: hydrogen peroxide and potassium permanganate. *Journal of Natural Fibers*, 2021 (Article in Press). https://doi.org/10.1080/15440478.2020.1870609.
27. Ivanovska A., Pavun L., Dojčinović B., Kostić M.: Kinetic and isotherm studies for the biosorption of nickel ions by jute fabrics. *Journal of the Serbian Chemical Society*, Vol. 86, No. 9, pp. 885-897, 2021. https://doi.org/10.2298/JSC210209030I.
28. Ivanovska A., Veljović S., Dojčinović B., Tadić N., Mihajlovski K., Natić M., Kostić M., A strategy to revalue a wood waste for simultaneous cadmium removal and wastewater disinfection. *Adsorption Science and Technology*, Vol. 2021, Article ID: 3552300, 2021. https://doi.org/10.1155/2021/3552300.
29. Mongioví C., Morin-Crini N., Lacalamita D., Bradu C., Raschetti M., Placet V., Ribeiro A.R.L., Ivanovska A., Kostić M., Crini G.: Biosorbents from plant fibers of hemp and flax for metal removal: Comparison of their biosorption properties. *Molecules*, Vol. 26, No. 14, 2021, Article No. 4199. https://doi.org/10.3390/molecules26144199.
30. Mongioví C., Lacalamita D., Morin-Crini N., Gabrion X., Ivanovska A., Sala F., Placet V., Rizzi V., Gubitosa J., Mesto E., Ribeiro A.R.L., Fini P., De Vietro N., Schingaro E., Kostić M., Cosentino C., Cosma P., Bradu C., Chanet G., Crini G.: Use of chènevotte, a valuable co-product of

industrial hemp fiber, as adsorbent for pollutant removal. Part I: Chemical, microscopic, spectroscopic and thermogravimetric characterization of raw and modified samples. *Molecules,* Vol. 26, No. 15, 2021, Article No. 4574. https://doi.org/10.3390/molecules26154574.
31. Ivanovska A., Veljović S., Reljić M., Ladarević J., Pavun L., Natić M, Kostić M: Closing the loop: Dyeing and adsorption potential of mulberry wood waste, *Journal of Natural Fibers,* 2021 (Article in Press). https://doi.org/10.1080/15440478.2021.2009398.
32. Milanovic J.Z., Milosevic M., Korica M., Jankovic-Castvan I., Kostic M.M., Oxidized hemp fibers with simultaneously increased capillarity and reduced moisture sorption as suitable textile material for advanced application in sportswear. *Fibers and Polymers,* 22(7), 2021, pp. 2052-2062. https://doi.org/10.1007/s12221-021-0450-y.
33. Kramar A., Petrović, M., Mihajlovski, K., Mandić B., Vuković G., Blagojević S., Kostić M., Selected aromatic plants extracts as an antimicrobial and antioxidant finish for cellulose fabric- direct impregnation method. *Fibers and Polymers,* 22, 2021, pp. 3317–3325. https://doi.org/10.1007/s12221-021-3007-1.
34. Zdujić A., Trivunac, K., Pejić, B., Vukčević M., Kostić M., Milivojević M., A Comparative study of Ni (II) removal from aqueous solutions on ca-alginate beads and alginate-impregnated hemp fibers. *Fibers and Polymers,* 22(1), 2021, pp. 9-18. https://doi.org/10.1007/s12221-021-9814-6.
35. Knezevic M., Kramar A., Hajnrih T., Korica M., Nikolic T., Zekic A., Kostic M., Influence of potassium permanganate oxidation on structure and properties of cotton. *Journal of Natural Fibers,* Vo. 19, 2022, pp. 403–415. https://doi.org/10.1080/15440478.2020.1745120.
36. Korica M., Peršin Z., Fras Zemljič L., Mihajlovski K., Dojčinović B., Trifunović S., Vesel A., Nikolić T., Kostić M.M., Chitosan nanoparticles functionalized viscose fabrics as potentially durable antibacterial medical textiles. *Materials,* 14/13, 2021, Article No. 3762, https://doi.org/10.3390/ma14133762.
37. Dimić-Mišić K, Kostić M, Obradović B, Kuraica M, Kramar A, Imani M, Gane P.. Iso- and anisotropic etching of micro nanofibrillated cellulose films by sequential oxygen and nitrogen gas plasma exposure for tunable wettability on crystalline and amorphous regions. *Materials,* 14(13), 2021, Article No. 3571, https://doi.org/10.3390/ma14133571.
38. Asanovic A.K., Ivanovska M.A., Jankoska Z.M., Bukhonka N., Mihailovic T., Kostic M.M.: Influence of pilling on the quality of flax

single jersey knitted fabrics. *Journal of Engineered Fibers and Fabrics.* Vol. 17, 2022, pp. 1-13. https://doi.org/10.1177/15589250221091267.
39. Ivanovska A., Branković I., Lađarević J., Pavun L., Kostic M.: Oxidized jute as a valuable adsorbent for Congo Red from an aqueous solution. *Journal of Engineered Fibers and Fabrics.* Vol. 17, 2022, pp. 1-9. https://doi.org/10.1177/15589250221101380.
40. Mongioví C., Lacalamita D., Morin-Crini N., Gabrion X., Placet V., Lado Ribeiro A.R., Ivanovska A., Kostić M., Bradu C., Staelensg J-N., Martel B, Raschetti M., Crini G.: Use of chènevotte, a valuable co-product of industrial hemp fiber, as adsorbent for copper ions: Kinetic studies and modeling. *Arabian Journal of Chemistry,* Vol. 15, No. 4, 2022, Article ID 103742, https://doi.org/10.1016/j.arabjc.2022.103742.
41. Mongioví C., Crini G., Gabrion X., Placet V., Blondeau-Patissier V., Krystianiak A., Durand S., Beaugrand J., Dorlando A., Rivard C., Gautier L., Lado Ribeiro A.R., Lacalamita D., Martel B., Staelens J-N., Ivanovska A., Kostić M., Heintz O., Bradu C., Raschetti M., Morin-Crini N., Revealing the adsorption mechanism of copper on hemp-based materials through EDX, nano-CT, XPS, FTIR, Raman, and XANES characterization techniques. *Chemical Engineering Journal Advances,* 10, 2022, Article No. 100282. https://doi.org/10.1016/j.ceja.2022.100282.
42. C. Mongioví, N. Morin-Crini, V. Placet, C. Bradu, ARL. Ribeiro, A. Ivanovska, M. Kostić, B. Martel, C. Cosentino, G. Torri, V. Rizzi, J. Gubitosa, P. Fini, P. Cosma, E. Lichtfouse, D. Lacalamita, E. Mesto, E. Schingaro, N. De Vietro, G. Crini. Hemp-based materials for applications in wastewater treatment by biosorption-oriented processes: A review. In: *Cannabis/Hemp for Sustainable Agriculture and Materials.* D.C. Agrawal, R. Kumar and M. Dhanasekaran, eds. Springer Nature Singapore Pte Ltd., chapter 9, pp. 239-295, 2022. https://doi.org/10.1007/978-981-16-8778-5_9.
43. Ivanovska A., Maletić S., Djokić V., Tadić N., Kostić M.: Effect of chemical modifications and coating with Cu-based nanoparticles on the electro-physical properties of jute fabrics in a condition of high humidity. *Industrial Crops and Products,* Vol. 180, 2022, Article number: 114792. https://doi.org/10.1016/j.indcrop.2022.114792.
44. Ivanovska A., Lađarević J., Asanović K., Barać N., Mihajlovski K., Kostić M., Mangovska B.: Quality of cotton and cotton/elastane single jersey knitted fabrics before and after softening and in situ synthesis of Cu-based nanoparticles. *Journal of Natural Fibers* (Article in Press) 2022. https://doi.org/10.1080/15440478.2022.2070328.

45. Kostic, M., Imani, M., Ivanovska, A., Radojevic, V., Dimic-Misic, K., Barac, N., Stojanovic, D.,Janackovic, Dj., Uskokovic, P., Barcelo E., Gane, P.: Extending waste paper, cellulose and filler use beyond recycling by entering the circular economy creating cellulose-$CaCO_3$ composites reconstituted from ionic liquid. *Cellulose,* Vol. 29, No. 8, 2022, pp. 5037-5059. https://doi.org/10.1007/s10570-022-04575-w.
46. Kukić D., Ivanovska A., Vasić V., Lađarević J., Kostić M., Šćiban M.: The overlooked potential of raspberry canes: from waste to an efficient low-cost biosorbent for Cr(VI) ions. *Biomass Conversion and Biorefinery,* 2022, (Article in Press) https://doi.org/10.1007/s13399-022-02502-4.
47. Potthast A., Ahn K., Becker M., Eichinger T., Kostic M., Böhmdorfer S., Jeong M.J., Rosenau T., Acetylation of cellulose–Another pathway of natural cellulose aging during library storage of books and papers. *Carbohydrate Polymers,* 287, 2022, Article No. 119323. https://doi.org/10.1016/j.carbpol.2022.119323.
48. Imani M., Dimic-Misic K., Kostic M., Barac N., Janackovic D., Uskokovic P., Ivanovska A., Lahti J., Barcelo E., Gane P.: Achieving a superhydrophobic, moisture, oil and gas barrier film using a regenerated cellulose-calcium carbonate composite derived from paper components or waste. *Sustainability,* Vol. 14, No. 16, 2022, Article number: 10425 2022. https://doi.org/10.3390/su141610425.
49. Ivanovska A., Asanović K., Jankoska M., Pavlović S., Poparić G., Kostić M.: Alkali treated jute fabrics suitable for the production of inexpensive technical textiles. Accepted for Publication in *Fibers and Polymers.*
50. Ivanovska A., Lađarević J., Asanović K., Pavun L., Kostić M., Mangovska B.: Revalorization of cotton and cotton/elastane knitted fabric waste. Accepted for publication in *Fibers and Polymers.*

Chapter 3

Progress, Challenges, and Prospects of Jute Fiber as Green Adsorbents: A Scope Beyond Traditional Applications

Aparna Roy[*], PhD
Department of Chemistry, Presidency University, Bengaluru, Karnataka, India

Abstract

Currently, water crisis and pollution and its management and possible solutions are recognized as a distinctive challenge faced by humankind. The major contributors to water pollution are discharging of industrial effluents in water bodies, contaminated with different heavy metal ions, dyes, hydrocarbons and other harmful chemicals. Though different techniques including photocatalytic decolorization and oxidation, biological degradation, coagulation and precipitation, ion exchange, membrane filtration, etc. are the conventionally used pollutant removal procedures, adsorption is one of the most popular and fundamental processes for wastewater treatment and water reclamation. Nowadays, among a large variety of adsorbents, activated carbon is one of the commericialized adsorbent for the wastewater treatment. But the adsorbent grade activated carbons are rather expensive and its usages are also associated with the difficulties of subsequent treatment, regeneration and disposal of the spent carbon. These constraints rendered the researchers to find a simple, economic and efficient adsorbent for pollutant removal from wastewater. From the view of environmental issues, in the search of alternative and inexpensive adsorbent, the efforts were mainly focused on the biological materials. Several experiments were conducted with a wide variety of biomaterials to investigate their

---

[*] Corresponding Author's Email: aparna.roy2006@gmail.com.

In: Jute: Cultivation, Properties and Uses
Editor: Matthicu Issa
ISBN: 979-8-88697-490-4
© 2023 Nova Science Publishers, Inc.

feasibility as adsorbents. However, most of these are unavailable plentifully in the global market, which eventually makes them incapable to meet the huge commercial demand, and maximum are also not highly efficient enough to be applied to real industrial wastewater. Thus, the exploration of a novel, environment friendlier, easily available, cheap and effective adsorbent for wastewater treatment, is still necessary. Regarding the current scenario, this review explores the feasibility of the novel application of abundantly available lignocellulosic jute fiber as a potential bioadsorbent for wastewater. Jute, the second most abundantly available natural bast fiber, is primarily comprised of cellulose (64.4%), hemicellulose (12%) and lignin (11.8%). Today, jute industry, a vital sector of South-East Asia, is critically challenged with several major threats to survival. Thus, widening the scope of jute fiber by utilizing it as bioadsorbent may offer a sustainable technology for wastewater treatment as well as promote the jute industries, which will be in turn beneficial for the jute farmers.

Keywords: literature review, jute fiber, adsorbent, wastewater, biomaterials

Introduction

The mankind with the progress in society, science, and technology has compelled the earth to face a serious water crisis. Water is one of the basic requirements for day to day survival of living beings, both in quantitative and qualitative terms (Toor, 2010). The consequences of the ever-growing population, rapid urbanization and industrialization, and improper utilization of natural water resources not only enhanced domestic and industrial water demands but also created environmental disorder by polluting fresh water. Currently, approximately 1.2 billion people are affected by the deficiency of safe and affordable water. Still now, the groundwater remained as sole source of drinking water in many rural areas and few large cities of some developing countries (Rijsberman, 2006). The economic prosperity, social stability, and aquatic systems provided by the resilience of ecological services could be threatened by the human impacts on the quality and quantity of fresh water (Ramachandra et al., 2002). The global water scarcity analyses estimated that two-thirds of the world population will be affected by water shortage over the next several decades (Yang et al., 2003). Hence, appropriate use of fresh water as well as reclamation, recycling and reuse of wastewater produced from industrial, agricultural and domestic activities is essential to abate the problem of water scarcity. UNESCO reported that the generation of global wastewater by industrial wastes and chemicals, domestic waste and agricultural wastes (fertilizers, pesticides and pesticide residues) is about 1,500 km^3, 1 L of which pollutes 8 L of freshwater and thus the present burden of pollution may be upto 12,000 km^3 worldwide (UNESCO, 2003). Thus, inadequately treated industrial wastewater may pose a chronic threat to the surface and groundwater resources by continuous discharge of effluents into them. The water is considered unsafe or too degraded to be used for domestic or industrial purposes, if the level of a pollutant in the water source exceeds an acceptable level. Hence, the focus on the reduction of pollution at the source and/or treatment of the polluted water prior to use were the best possible solutions to abide the pollution problems of aquatic ecosystems. The major water pollutants that seriously affect the biodiversity, ecosystem functioning, and the natural services of aquatic systems upon which society depend, include a variety of hazardous organic and inorganic materials (Ahalya and Ramachandra, 2002).

Over the last few decades, the wastewater treatment processes received considerable attention from researchers and several methods were devised for the removal of pollutants. The conventionally used procedures for pollutant removal from aqueous streams include adsorption, photocatalytic decolorization and oxidation techniques, biological degradation, coagulation and precipitation, ion exchange, membrane filtration etc. Among these different conventional procedure, adsorption, an integral to a broad spectrum of physical, biological, and chemical methods and operations in the environmental field, is one of the most popular and fundamental processes for wastewater treatment and water reclamation. Basically adsorption is a mass transfer phenomena where substances are accumulated at the fluid-solid interface by physical and/or chemical attractive forces. The proper design of the well known equilibrium separation process, adsorption, can economically meet today's higher effluent standards and water reuse requirements. The advantages, which made adsorption superior to the other techniques for water treatment, can be summarized as: (i) operate independently and effectively, (ii) flexible and simple in design, (iii) operation and maintenance of the process is safe, reliable, and easy, no skilled operators are needed, (iv) minimum energy consumption, (v) the efficiency is not decreased drastically in presence of competing ions of different salts like, sulfate, nitrate, chloride etc., (vi) user friendly at the household level and it has well social acceptability, (vii) cost effective in terms of set-up, operation and maintenance. The overall process cost mainly depends on cost of adsorbent itself. Cheaper the adsorbent cost, the more inexpensive is the process.

According to the abundant literature data, several low cost new adsorbents were already studied and there are endless opportunities to develop new ones (Maiti, 2010).

In this context, jute, a lignocellulosic and ecofriendly bast fiber, can also be considered as a potential candidate for bioadsorption. Jute is extracted from two herbaceous annual plants, viz., white *Corchorus capsularis* (white jute) and *Corchorus olitorius* (Tossa jute). It is composed of 58–63% cellulose, 20–22% hemicellulose, 12–15% lignin, approximately 2% protein, 1% mineral, trace quantities of organic and inorganic pigments. Next to cotton, in terms of production and use, jute is the second most common natural fiber, cultivated in the world and extensively grown in India, Bangladesh, Indonesia, and China. The Indian jute industry plays a pivotal role in the country's economy. The inherent characteristics, like, strength, versatility, renewability, eco-friendliness, biodegradability, etc., of jute fiber directly catered to technical requirements and received increasing attention from industry for its appropriate end applications. Several technological developments not only unrestricted the limitations of jute as a traditional textile fiber, but also enabled to extend its arena as a raw material for manufacturing other value-added products such as, pulp and paper, geo-textiles, composites and home textiles etc., which are extremely attractive in the view of environmental and ecological awareness. India's share in global exports of reusable, sustainable and biodegradable jute goods is around 25%. Today, jute industry, a vital sector of India, is confronted with many problems. Thus, widening the scope of jute fiber by utilizing it as bioadsorbent may offer a sustainable technology for pollutant removal as well as promote the jute industries, which will be in turn beneficial for the jute farmers. The worldwide support towards environmental commitments, created a systematic drive to raise the production and productivity of this crop so as to provide new opportunities for jute along with the socio-economic welfare of the jute farmers (Kumari et al., 2018).

Thus, with the aim of exploring the efficacy of jute fiber as a novel, feasible, and cost-effective adsorbent, several researchers studied the aptitude of the untreated and treated jute fiber for biosorption of different hazardous pollutants such as, heavy metals (Cr^{+6}, Cu^{+2}, Ni^{+2}, Zn^{+2} etc.), dyes (congo red, methylene blue, acid violet, eosin yellow, malachite green, brilliant blue, crystal violet, etc.), and other organics (aniline, oils), present in wastewaters.

Adsorption of Inorganic Heavy Metal Ions by Untreated/Treated Jute Fiber

Shukla and Pai (2005), studied the potential of untreated and treated jute fiber for adsorption of three heavy metal ions viz., Cu(II), Ni(II) and Zn(II) from their aqueous solutions. Chemical modification of jute fiber was carried out by Reactive Orange 13 dye and hydrogen peroxide. Though both the modified jute fiber showed better efficacy for adsorption of the heavy metal ions than raw jute fiber, it was observed that dye loaded jute fiber had lower metal adsorption capacity than the hydrogen peroxide oxidized fiber. The metal uptake decreased with decreasing the pH of the solution. The adsorption of metal ions by modified jute fiber fits Langmuir adsorption isotherm model. Desorption of the spent adsorbent was carried out by hydrochloric acid and nitric acid, and the regeneration was done by treating the desorbed fiber with NaOH solution. It was observed that desorption is sufficiently high for unmodified jute whereas with modified jute it is nearly complete.

Kumar et al., (2008) prepared chemically modified jute fiber based adsorbent by synthesizing short chain of polyaniline (PANI) (oligoaniline) on the fiber surface. Aniline, ammonium peroxydisulfate (oxidant), 1,4-phenylenediamine (chain terminating agent), and raw jute fiber were mixed together in acidic aqueous medium at 5 °C and the reaction mixture was kept overnight. The PANI-jute fiber was obtained after decanting the liquid. PANI-grafted jute fiber was utilized for adsorption of Cr(VI) from its aqueous solution in batch experiment as well as fixed bed column studies. The effect of different variables, viz., chain length of PANI on fiber surface, solution pH, initial concentration of Cr(VI), dose of PANI-modified jute and temperature, were evaluated in batch mode. The experimental data was analyzed with Langmuir and Freundlich adsorption isotherm models and maximum monolayer capacity of PANI-modified adsorbent was calculated to be 62.9 mg/g at pH 3 and 20° C temperature. The chromium adsorption by chemically modified jute fiber was exothermic in nature, since removal of metal was observed to decrease with rise in temperature. 83% desorption was achieved by treating the spent adsorbent by NaOH (sodium hydroxide, 2M) for 10 min.

In continuation of the aforementioned studies, Kumar and Chakraborty (2009) also conducted fixed-bed column experiments. The variable parameters chosen for evaluation of Cr(VI) adsorption from aqueous solution by PANI-jute fiber in fixed-bed column mode were influent pH, column bed depth, influent Cr(VI) concentrations and influent flow rate. Similar to batch adsorption studies, the maximum removal of total chromium was achieved at pH 3. As the column bed depth increased from 40 to 60 cm, both total chromium uptake and throughput volume at exhaustion point also increased. Authors observed a nice correlation between theoretical and experimental breakthrough profile till 10% breakthrough by applying BDST (Bed Depth Service Time) equation for the collected batch data incorporated into Langmuir isotherm model.

A low-cost, environment-friendly and recyclable biosorbent derived from jute fiber was developed by Du et al., (2016) for efficient adsorption of Pb(II), Cd(II) and Cu(II) ions from aqueous solutions. For surface functionalization of jute fiber, they adopted a fast and facile method. Initially, the raw jute fiber was pretreated with alkaline (NaOH) solution under microwave heating. In the next step, rapid grafting of carboxylic groups on the pretreated jute surface was achieved via esterification reaction by continuously stirring the mixture of pyromellitic dianhydride (dissolved in dimethylformamide) and virgin jute into the microwave reactor for 15 min at 393 K temperature. The chemical and morphological changes of the prepared bioadsorbent were investigated by FTIR (Fourier transform infrared), SEM (scanning electron microscope), and EDX (energy-dispersive X-ray) characterization techniques. The spontaneous and endothermic adsorption behavior of carboxyl-modified jute fiber towards the heavy metal ions, followed pseudo second-order kinetics model and Langmuir isotherm model ($R^2 > 0.99$) with higher adsorption capacity than raw jute. Also, the spent adsorbent was regenerated with Ethylenediamine tetraacetic acid (EDTA) solution and reused up to at least four times with equivalent high adsorption capacity. The carboxyl modified jute fiber was also utilized for Cu (II) removal from aqueous solution, by means of fixed bed column adsorption (Du et al., 2018). The effects of adsorbent dosage, flow rate and the initial Cu(II) concentration on the corresponding breakthrough curves and dynamic adsorption performance of raw and modified jute were examined. Along with the amount of Cu(II) removed, the breakthrough time increased as the adsorbent dosage was increased, whereas the superficial flow rate and initial Cu(II) concentration decreased. The adsorption efficiency of Cu(II) by fixed bed adsorption process with untreated and esterified jute fiber was further described by Thomas, Yoon-Nelson, and BDST models ($R^2 > 0.90$).

Instead of directly employing jute fiber as adsorbent, Rahman et al., (2020), extracted pure cellulose from jute fiber by boiling and bleaching it with sodium hydroxide solution (15%) and hydrogen peroxide (50%), respectively. Extracted cellulose was converted to poly(methyl acrylate)-grafted and poly(acrylonitrile)-grafted cellulose by applying free radical grafting chemistry, followed by subsequent conversion to poly(hydroxamic acid) and poly(amidoxime) ligands. The prepared cellulose derivative was then utilized as an adsorbent, which efficiently removed 98% of Cu(II) and approximately 90% of Co(II), Cr(II), Ni(II), and Pb(II) ions from electroplating wastewater. The multiple-layers adsorption showed good correlation with Freundlich isotherm model.

From the perspective of expanding the lifecycle of adsorbents to minimize the waste accumulation and feedstock consumption, Huang et al., (2019) designed sequential remedial of aniline and heavy metal ions by jute fiber-based biosorbents. The achievement of this successive removal strategy depends on the fundamentals of the surface chemistry of adsorbents. The pristine jute fibers were first used for adsorbing highly toxic aniline from water. The reactivity of the surface adsorbed aniline was then reasonably utilized for chemical modification of jute fiber. The surface bound aniline was polymerized using ammonium persulfate initiator at 5 °C via in-situ oxidation to form polyaniline grafted jute fiber. This simple and practical chemical conversion not only saved the trouble of desorption, but also endowed the spent adsorbent with a new purpose of removing heavy metal ions (Cd^{+2}, Cr^{+6}) by converting it into a better functionalized material. The conjugated structure and redox properties of PANI molecules enabled PANI coated jute surface to scavenge and decontaminate toxic metal ions. Overall, this investigation successfully contributed towards the development of a sustainable water treatment technology by demonstrating a unique approach for adsorbent regeneration and waste disposal.

Adsorption Mechanism of Heavy Metal Ions by Jute Fiber Based Adsorbent

Adsorption mechanism largely depends on the physical and chemical characteristics of both the adsorbent and the adsorbate (Roy et al., 2013a). Based on experimental results and spectroscopic revelation of the presence of various functional groups on the surface of jute fiber based adsorbent, and considering the positive charge of the heavy metal ions it was inferred that the main mechanism of heavy metal ions removal was the ion exchange (Du et al., 2016). The adsorption of heavy metal ions by jute fiber based adsorbent may also occur by electrostatic attraction or complexation.

Adsorption of Dyes and Other Organic Pollutants by Untreated/Treated Jute Fiber

Plethora of literature are available regarding the removal of different types of hazardous acidic, basic or reactive dyes with jute fiber based bioadsorbents.

Jute processing waste (raw as well as pyrolysed under nitrogen and steam atmospheres), generated during the cultivation and processing of jute fibers (including unused jute fibers, cut off stocks and reject fibers of inferior grade), were employed for the treatment of aqueous solutions of most commonly used dyes, like methylene blue, congo red and acid violet. The batch adsorption studies revealed that raw jute processing waste was a better material for treating colored water containing synthetic dyes compared to its powdered and granular activated carbon form. The jute processing waste based adsorbent was further utilized for biological oxygen demand (BOD) and chemical oxygen demand (COD) removal from jute retting effluent and the raw one was found to have less potential as compared with its granular activated carbon (GAC) form (Banerjee and Dastidar, 2005).

Senthilkumaar et al., (2005) prepared activated carbon from jute fiber using phosphoric acid for adsorption of methylene blue from its aqueous solution. Sun dried jute fiber was carbonized with 15% phosphoric acid and employed for batch adsorption of another acidic dye, eosin yellow (Porkodi and Kumar, 2007) from its aqueous solution. Apart from the removal of acidic dye, Porkodi and Kumar (2007), also utilized the carbonized jute fiber for adsorption of other two basic dyes, viz., malachite green and crystal violet. The process variables studied are, initial dye concentration, initial solution pH, adsorbent dosage, and contact time. The adsorption of eosin yellow favored lower pH and Langmuir isotherm model, whereas adsorption of malachite green and crystal violet prefers higher pH and Freundlich isotherm model. The pseudo second order model successfully described the kinetics of the studied dyes onto carbonized jute fiber based adsorbent.

However, Congo Red was chosen as model acidic dye by Roy et al., (2012a) to study the feasibility of jute fiber as a potential adsorbent for removal of textile dyes. Both batch and fixed bed column experiments were performed for adsorption of Congo Red by untreated and surface modified jute fiber. Authors introduced a natural polyphenol containing tannin onto the fiber surface by covalent grafting to further improve the adsorption efficiency of jute fiber. Successful grafting of tannin onto jute fiber was achieved after pretreatment of the jute surface with sodium hydroxide followed by functionalization of the jute surface with epoxy functional group using epichlorohydrin (Roy et al., 2013a, b). The isotherm and kinetic investigations based on the experimental results for both the batch and column experiments were further correlated to establish that Langmuir model and pseudo-second-order model were the best fit as confirmed by high correlation coefficients. Desorption of the spent adsorbent was carried out by 0.1 M NaOH solution and the optimum conditions for this decolorization were deduced by using a full factorial central composite design (CCD) in response surface methodology (RSM). This applied methodology found to be very promising to achieve maximum color removal from the real textile dyeing effluent (Roy 2021).

Jute fiber based adsorbent was also used for decontamination of reactive dyes like Remazol Brilliant Blue BB and Reactive Red 195. For adsorption of Remazol Brilliant Blue BB, stepwise chemical treatment was done on the jute fiber surface. First, jute fibers swelled in NaOH solution (20%) for 24 h, followed by grafting with acrylic acid by gamma irradiation technique. The acrylic acid grafted fibers were then immobilized with chitosan. The successful chemical modification of jute fiber was verified by FTIR, SEM, XRD (X-Ray diffraction), and TGA (Thermogravimetric analysis). Remazol Brilliant Blue BB removal by the chitosan immobilized acrylic acid grafted jute fiber was demonstrated using Langmuir and Freundlich equilibrium isotherm models (Hassan, 2015). On other hand, the entropy driven, endothermic adsorption of reactive red 195 by NaOH pretreated jute fiber was optimized using face centered response surface methodology (RSM) and it was found that optimum adsorption capacity can be achieved at neutral pH, low to moderate temperature, moderate to high jute dose and moderate RPM. Alkali treatment of pristine jute fiber was done to enhance its adsorption performance (Dey and Dey, 2021). During treating the jute fiber with aqueous sodium hydroxide solution, amorphous lignin and hemicellulose leaches out, and thus makes the fiber surface more active (Roy et al., 2012b).

Dey et al., (2022) reported optimization of the removal of an acidic azo dye, methyl red, from aqueous solution applying sodium carbonate treated jute fiber as adsorbent by response surface methodology approach. Rotational speed, temperature, pH, and adsorbent dose were chosen as independent variable parameters for maximum adsorption capacity during dye removal.

Removal of other hazardous organic chemicals, like aniline and oil, by jute fiber based adsorbents were studied by Hu et al., (2015, 2016) and Lv et al., (2018). The virgin jute fiber was treated with pyromellitic dianhydride (dissolved in N, N-Dimethyl formamide solvent), under microwave treatment at 123 °C for 25 minutes and the prepared pyromellitic dianhydride modified jute fiber was employed as biosorbent for the decontamination of aniline, both in batch as well as column adsorption modes. Various factors influencing the adsorption behavior e.g., initial pH of solution, initial concentration of adsorbate, adsorbent dosage and temperature were studied and it was revealed that the adsorption was spontaneous and endothermic in nature and follows the pseudo second order kinetic model and Langmuir isotherm model. The spent adsorbent was also regenerated through desorption of aniline using 0.5 M HCl solution Hu et al. (2015, 2016). For adsorption of a light diesel oil, containing paraffins, naphthenes, and aromatic hydrocarbons of 9–18 carbon atoms, Lv et al., (2018) prepared a unique jute fiber based hydrophobic oil adsorbent. The pristine jute fiber were initially pulverized and pretreated with NaOH solution. In the second step, a layer of silica (SiO_2) particles was incorporated onto the surface of the alkali treated jute fiber, applying ethyl orthosilicate via sol-gel method. Finally, hydrophobic alteration of the adsorbent was achieved by immersing it into octadecyltrichlorosilane solution. The batch adsorption data of diesel by modified jute fiber demonstrated that the process is spontaneous, favorable, and exothermic. The obtained experimental data had good correlations with Freundlich isotherm and pseudo-second-order kinetic models. Sufficient buoyancy, biodegradability, and environment friendliness, made the hydrophobically modified jute fiber suitable for marine oil spill cleanup (Lv et al., 2018).

Adsorption Mechanism of Heavy Metal Ions by Jute Fiber Based Adsorbent

The plausible mechanism of adsorption of dyes and other organic contaminants onto jute fiber based adsorbent can be deduced on the basis of physical or chemical characteristics of adsorbents and adsorbates. Different spectroscopic characterization of jute fiber based adsorbent confirmed the major presence of polar functional groups, which indicates the possibilities of formation of physical bonds via hydrogen bonding or van der Waals forces between the organic adsorbate molecules and jute fiber surface, during adsorption process. Hence, physic-adsorption may be a plausible mechanism of organic pollutants removal by jute fiber based adsorbents (Roy et al., 2013a).

Challenges of Jute Fiber as Adsorbent

Indeed jute fiber gained increased interest to be applied in the field of water purification technology and substantially explored by scientific community across the globe to validate their feasibility as a potential bio-adsorbent for inorganic and organic pollutant. It's large and porous surface along with multiple functionalities including carboxylic and hydroxylic groups, enabled them to interact with the adsorbates via different forces e.g., electrostatic, Van der Waals interactions, H-bonding, etc., to make them a candidate of interest as bio-adsorbent. However, there are certain limitations as well in terms of availability, efficiency, consistency in performance, very low resistance to microbial attack, scaling up for wide application, disposal of spent jute fibers etc.

Nevertheless, jute fiber is very cheap and second most abundantly available fiber, global production of jute fiber is merely 2,688,912 tonnes and the only countries that produces significant quantities of jute fiber are India, Bangladesh, and China (Kumari et al., 2018). Thus, the insufficiency to meet huge commercial demand in the global market can be considered as one of the challenges to apply jute fiber as adsorbent.

Moreover, like any other lignocellulosic biomass, the adsorption capacity of jute fiber varies greatly according to their origin, type and region of cultivation. Also, due to their low specific surface area, they are not highly efficient enough in their virgin form, to be applied for removal of different contaminants, compared to commercially available adsorbents (such as activated carbon, silica gel, alumina) (Roy et al., 2012a). Another major drawback associated with the jute fiber is their very low resistance to microbial attack, which reduces its stability. The obstacles regarding inefficiency and instability, were tried to vanquish by adopting different chemical modifications through incorporating new functionalities to cellulose backbone. Nonetheless, the major flaw of most of these functionalization methodologies is the use of hazardous chemicals and/ or organic solvents, inconsistency in extent of chemical modification, and afterwards possibility of leaching of chemicals treated with.

Other aspects behind the inappropriateness of jute fiber are desorption, regeneration and disposal of spent adsorbent. Investigations conducted earlier demonstrated that desorption efficiency of jute based adsorbent were poor and a significant amount of contaminant could not be desorbed. This indicated that the disposal of used adsorbents is more cost-effective than its regeneration.

But safe disposal of the spent jute fibers require a serious assessment of their possible toxicity and effects towards the environment (Roy et al., 2013a).

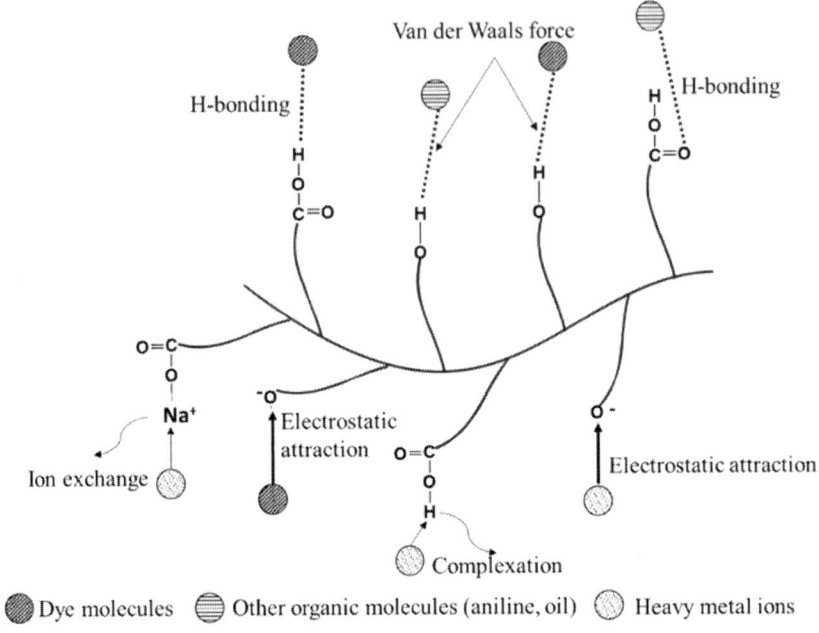

Figure 1. Adsorption of inorganic and organic contaminants by jute fiber based adsorbent.

Since, jute fiber as adsorbent is not yet a fully commercially available commodity, another concern associated with it, is scaling–up the process for easy adaptation by the industries. Though all the researchers observed promising results in laboratory scale, the investigation had not been carried out further to validate at larger scale. In fact, except few, most of the adsorption studies were performed with synthetic wastewater instead of the real sample. So, pilot scale experiments are needed to be designed to demonstrate whether jute fiber based adsorbents are efficient enough to be applied for outdoor performance under the real scenario.

Conclusion

Direct discharge of contaminated industrial effluents into the water resources pose a serious threat to the environment. Thus exploitation of safe water sources is a global challenge that requires the reuse of reclaimed wastewater for sustainable development to save planet and mankind. Among several available techniques for the removal of toxic pollutants, adsorption is the most preferred and widely accepted one owing to its simplicity, ease of operation, high efficiency and economic feasibility.

Major outcome from the studies discussed above can be emphasized as inexpensive development of jute fiber based adsorbent and exploration of its potential as efficient biosorbent for removal of different heavy metal ions (Cr^{+6}, Ni^{+2}, Cu^{+2}, Pb^{+2}, Zn^{+2}, Cd^{+2}), dyes (methyl red, methylene blue, congo red, eosin yellow, crystal violet etc.) and other organic pollutants (aniline and oil). The operating parameters of adsorption process (pH, adsorbent dose, initial dye concentration, and temperature) onto jute fiber based adsorbent had strong influence on their adsorption efficiency. Batch adsorption studies were spontaneous and attained rapid equilibrium which is desirable and advantageous for practical adsorption applications. However, as estimated from adsorption isotherm studies, the maximum adsorption capacity of the jute fiber based adsorbents were not very high to promote this solution towards implementation as a competent commercial adsorbents for water filtration. Typically, the adsorption kinetic data for pollutant removal by untreated and treated jute fiber favored pseudo-second-order kinetic model. Mostly, desorption and regeneration of spent adsorbent was not efficient and cost-effective. The column adsorption study revealed that better column efficiency, with slower saturation of column, was achieved at longer bed depth, lower flow rate, and diluted inlet dye concentration. The forces, e.g., electrostatic interaction, hydrogen bond, van der Waals force etc., between the largely available functional groups of jute fiber based adsorbent and the pollutants, make the adsorption process feasible. Nevertheless, jute fiber based adsorbent have been proven moderately effective in treating waste water with their economical feedstock and ease in preparation process. Few concerns like bulk scale production of jute fiber based adsorbent, designing of pilot-scale continuous-flow filter for its practical applicability, further improvement of adsorption capacity and selectivity for removal of different anionic, cationic and anionic/cationic species from liquid phase etc., needed to be meticulously addressed. Thus, the applicability of jute fiber may be broadened as a potential, alternative and low-cost adsorbent to introduce a new horizon for further advances to purify the wastewater, which is currently one of the most challenging global issue against more sustainable planet.

Disclaimer

None.

References

Ahalya N, Ramachandra TV. Restoration of wetlands - Feasibility Aspects of Biological Restoration. *National Conference on Aquatic Restoration and Biodiversity*, India; 2002.

Banerjee S, Dastidar MG. Use of jute processing wastes for treatment of wastewater contaminated with dye and other organics. *Bioresource Technology* (2005) 96:1919-1928.

Dey AK, Dey A. Selection of optimal processing condition during removal of Reactive Red 195 by NaOH treated jute fibre using adsorption. *Groundwater for Sustainable Development* (2021) 12:100522.

Dey AK, Dey A, Goswami R. Adsorption characteristics of methyl red dye by Na_2CO_3-treated jute fibre using multi-criteria decision making approach. *Applied Water Science* (2022) 12:179-191.

Du Z, Zheng T, Wang P. Experimental and modelling studies on fixed bed adsorption for Cu(II) removal from aqueous solution by carboxyl modified jute fiber. *Powder Technology* (2018) 338:952-959.

Du Z, Zheng T, Wang P, Hao L, Wang Y. Fast microwave-assisted preparation of a low-cost and recyclable carboxyl modified lignocellulose-biomass jute fiber for enhanced heavy metal removal from water. *Bioresource Technology* (2016) 201:41-49.

Hassan MS. Removal of reactive dyes from textile wastewater by immobilized chitosan upon grafted Jute fibers with acrylic acid by gamma irradiation. *Radiation Physics and Chemistry* (2015) 115:55-61.

Hu Q, Pan H, Jiang J, Gao DW, Wang P. High-capacity adsorption of aniline using surface modification of lignocellulose-biomass jute fibers. *Bioresource Technology* (2015) 193:507-512.

Hu Q, Wang P, Jiang J, Pan H, Gao DW. Column adsorption of aniline by a surface modified jute fiber and its regeneration property. *Journal of Environmental Chemical Engineering* (2016) 4:2243-2249.

Huang Q, Hu D, Chen M, Bao C, Jin X. Sequential removal of aniline and heavy metal ions by jute fiber biosorbents: A practical design of modifying adsorbent with reactive adsorbate. *Journal of Molecular Liquids* (2019) 285:288-298.

Kumar PA, Chakraborty S. Fixed-bed column study for hexavalent chromium removal and recovery by short-chain polyaniline synthesized on jute fiber. *Journal of Hazardous Materials* (2009) 162:1086-1098.

Kumar PA, Chakraborty S, Ray M. Removal and recovery of chromium from wastewater using short chain polyaniline synthesized on jute fiber. *Chemical Engineering Journal* (2008) 141:130-140.

Kumari K, Devegowda SR, Kushwaha S. Trend analysis of area, production and productivity of jute in India. *The Pharma Innovation Journal* (2018) 7(12):58-62.

Lv N, Wang X, Peng S, Zhang H, Luo L. Study of the Kinetics and Equilibrium of the Adsorption of Oils onto Hydrophobic Jute Fiber Modified via the Sol-Gel Method. *International Journal of Environmental Research and Public Health* (2018) 15:969-983.

Maiti A. *Removal of arsenic from water using raw and treated laterite as adsorbent*, PhD Thesis, Indian Institute of Technology, Kharagpur, India; 2010.

Porkodi K, Kumar KV. Equilibrium, kinetics and mechanism modeling and simulation of basic and acid dyes sorption onto jute fiber carbon: Eosin yellow, malachite green and crystal violet single component systems. *Journal of Hazardous Materials* (2007) 143, 311-327.

Rahman ML, Fui CJ, Ting TX, Sarjadi MS, Arshad SE, Musta B. Polymer ligands derived from jute fiber for heavy metal removal from electroplating wastewater. *Polymers* (2020) 12:2521-2548.

Ramachandra TV, Kiran R, Ahalya N. Status, *Conservation and Management of Wetlands*. Allied Publishers (P) Ltd, India; 2002.

Rijsberman FR. Water scarcity: Fact or friction. *Agriculture and Water Management* (2006) 80:5-22.

Roy A. Removal of color from real textile dyeing effluent utilizing tannin immobilized jute fiber as biosorbent: Optimization with Response Surface Methodology. *Environmental Science and Pollution Research* (2021) 28(10):12011-12025.

Roy A, Chakraborty S, Kundu SP, Adhikari B, Majumder SB. Lignocellulosic jute fiber as a bioadsorbent for the removal of azo dye from its aqueous solution: batch and column studies. *Journal of applied polymer science* (2013a) 129:1-13.

Roy A, Chakraborty S, Kundu SP, Majumder SB, Adhikari B. Surface grafting of *Corchorus olitorius* fiber: a green approach for the development of activated bioadsorbent. *Carbohydrate Polymers* (2013b) 92(2):2118-2127.

Roy A, Adhikari B, Majumder SB. Equilibrium, kinetic, and thermodynamic studies of azo dye adsorption from aqueous solution by chemically modified lignocellulosic jute fiber. *Industrial & Engineering Chemistry Research* (2012a) 52:6502−6512.

Roy A, Chakraborty S, Kundu SP, Basak RK, Majumder SB, Adhikari B. Improvement in mechanical properties of jute fibers through mild alkali treatment as demonstrated by utilisation of the Weibull distribution model. *Bioresource Technology* (2012b) 107: 222-228.

Senthilkumaar S, Varadarajan PR, Porkodi K, Subbhuraam CV. Adsorption of methylene blue onto jute fiber carbon:kinetics and equilibrium studies. *Journal of Colloid and Interface Science* (2005) 284:78-82.

Shukla SR, Pai RS. Adsorption of Cu(II), Ni(II) and Zn(II) on modified jute fibres. *Bioresource Technology* (2005) 96:1430-1438.

Toor MK. *Enhancing adsorption capacity of Bentonite for dye removal: Physicochemical modification and characterization*, The University of Adelaide, Australia; 2010.

UNESCO. Water for People Water for Life, *The United Nations World Water Development Report*; 2003.

Yang H, Reichert P, Abbaspour K, Zehnder AJB. A water resources threshold and its implications for food security. *Environmental Science and Technology* (2003) 37:3048-3054.

Chapter 4

Jute Fiber: Extraction, Properties and Applications

Md. Vaseem Chavhan[1], B. Venkatesh[1] and Beera Murali[2]

[1]Department of Knitwear Design,
National Institute of Fashion Technology Hyderabad, India
[2]Department of Handlooms and Textiles,
Indian Institute of Handloom Technology Salem, India

Abstract

The jute fibre (*Corchorus olitorius*) belongs to the category of natural cellulosic bast fibre. The fibre cells are separated from the bast by the retting process to extract the individual fibres. The different media, moisture, water, chemical, and enzymes are used for the retting process based on the time of the process and the quality of fibre to be obtained. The extracted fibre is further processed to obtain the finished fibre of the required fineness and luster. Jute fibers are characterized by their unique multicellular structure, having cellulosic microfibrils covered with lignin. The jute fibre is known for its good initial modulus, dimensional stability, toughness, and antimicrobial properties among cellulosic natural fibres. Other than these properties, the low cost of fibre, and the world's focus towards sustainability, the jute fibre is gaining its scope in various applications like fibre reinforced composites, geotechnical, automotive, and construction applications. In the present chapter, the jute fibre cultivation, and extraction by retting have been discussed. Further, the conversion of extracted fibre to yarn and fabric is explained in detail.

Keywords: jute fibre, retting, spinning, lubrication, weaving

In: Jute: Cultivation, Properties and Uses
Editor: Matthieu Issa
ISBN: 979-8-88697-490-4
© 2023 Nova Science Publishers, Inc.

Introduction

According to the classification of textile fibers, cellulosic fibers are divided into three groups 1) Stem or bast fibers 2) Leaf fibers and 3) seed and fruit fibers. Jute is extracted from the bast or skin of the plant, so it falls under the bast category in fiber classification. Among all the bast fibers, jute tops first. Jute is one of the cheapest natural fibers. These fibers comprise 65-75% cellulosic and 3-10% lignin, 15-20% hemicellulose, 3% pectin water-soluble matter 1.5%, and fat and wax 0.6%, which are components of plant and wood. It is also known as lignocellulosic fiber. Jute is named after the fiber extracted from the stems of plants belonging to the genus *Corchorus*, family *Tiliaceae*. Due to the appearance of light golden-brown color, it is also termed "golden fiber". Jute can easily be recyclable and biodegradable due to its excellent characteristics; it is an environmentally friendly material.

Jute cultivation requires a warm and humid climate with a temperature range between 22°C to 35°C. It is an art of the Indian farmers. Drylands are used for the cultivation of jute for a period of 4 to 5 months. The water requirement for cultivation is significantly less, and it grows like a leafy plant to a height of 7 to 12 feet and a stack diameter of around 1 inch. The process of harvesting in India is by hand sickle. Then it is processed by different methods like retting, stripping, washing, and drying, in which the fibers are separated from the plants. These fibers have good insulating properties for thermal and acoustic energies with moderate moisture regain. Moisture plays a vital role in jute fiber. Under humid conditions, it can absorb a large quantity of water. Moisture is also one of the factors for the deterioration of jute fibers, affecting its strength and utility. These fibers also have less skin irritation. The annual production of jute fiber is around 4 million tons and has been used for various applications. Jute yarns are mainly used in making bags for storage, which is the biggest consumer of jute in the markets. Jute bags have gained an advantage as being eco-friendly instead of non-biodegradable materials.

Process of Fiber Extraction from Jute

The extraction process of jute fiber from the jute plant begins soon after the cutting of plants in the fields, which will be done in different stages and are shown in Figure 1.

Figure 1. Jute fiber extraction flow chart.

Bundle Stalk

Soon after the plant grew, pluck from the land along with the roots, and made a bunch in the fields, and later all the bunches were made into bundles. In few areas of India, they cut the grown plant at the bottom of the stem. Cutting instead of plucking is to eliminate the cutting process before loading the material in the spinning process. Pulling the stem along with the roots gives better fiber extraction from the jute, while in the retting process, the jute stem must be deep water for 3 to 5 days. In these conditions pulling is the best suitable than cutting from the fields.

Retting

Retting is an essential operation in jute fiber extraction. In this process, the fibers are separated from the non-fibrous and woody parts of the stem. The efficiency of the retting process determines the quality of fiber extracted. During the retting process, pectic materials break down, and fibers are formed. Retting time and conditions during the process determine the fiber quality. Usually, the bottom part of the jute plant is hard and thick, so it will take more time to retting compared to the upper part of the plant. So before retting the bottom part of the plant is cut, the remaining portion will be subjected to retting [1].

Process of Jute Retting

The Retting process involves the removal of non-fibrous content from the plant stem either by microbial method or by using chemicals where the fiber bundles are loosened and attached tissues are removed during the washing process. Fiber quality greatly depends on the process parameters maintained during the retting process [2].

Retting is a process in which the bundles of jute stalks are placed in a tank filled with water, where fibers get loosened and separated from the woody stem. Traditionally, retting is processed after harvesting jute bundles are brought to the retting site, and framers carry the process of retting in small ponds or lakes in their villages.

The bundles are dropped in lakes or ponds used to float on water. In this, the farmers placed one upon another in reverse and criss-cross directions. These bundles are tied with split bamboo on both sides to avoid them from drifting away. This process is called '*JAK*' which prevents it from drifting, and bottom poles are pushed through the jak into the bottom of the retting channel. Due to these floating conditions, they are covered with plant-based materials like paddy straws, and long leave grass to prevent the bundles from floating. In few cases long bamboo logs are used to place on the bundles. This helps the bundle to submerge in the water. The submerging is very important in the retting process; if any portion of the bundle is not submerged will impact the retting process. The retting process also depends upon the amount of nitrogen available.

The bark of the freshly plucked jute can be split loose by splitting fiber from the wood it surrounds. The extraction depends upon the age of the crop. The over-retted jute becomes fragile and loses luster. Extraction can be carried out only when most of the fiber is loosened. The wet stripes of fiber are stretched in water and washed thoroughly to remove gum and dirt remains from the decomposed plant. This is a critical process in few places due to the scarcity of water, this process is eliminated. This causes the dirt and dust has been carried along with the fiber for further processing. In tank retting process, bundles are steeped in water at least 60 cm to 100 cm depth. The Retting process time limit is decided according to the thickness or diameter of the wood stalk. Excess retting causes damage to the jute fiber. The completion of retting can be determined when the barks separate easily from the wood stalk and the fibers are ready for extraction.

After this process, the fibers are washed, cleaned thoroughly, and dried in the outside atmosphere under mild sunlight. The good quality of fibers can be linked to the accessibility of abundant water. The fibers are not to be dried on the ground and spread on a bamboo raft as wet fibers can easily catch dirt and dust on the floor. These dried fibers are bundled and sent for further processing.

Advantages of Jute Retting by Nature Bodies

- Artificial activities are not required as it is naturally retted.
- No extra labor is required
- Cost efficiency

The Important Conditions for the Retting Process

- The water must be non-saline and perfect.
- Sufficient capacity of water to be chosen to let the jute bundles float.
- Jute bundles should not touch the bottom when they are submerged.
- In the case of a tank retting, the same tank or ditch may not be used repetitively.

The Retting process can be done in two ways 1) wet Retting and 2) Dry retting, wet retting is discussed above, and dry retting is carried out when there is a lack of water after the harvesting of the plants are allowed to dry. The bundled jute plants are placed longitudinally and allowed to dry this dry retting does not give profitable results in extraction of jute fibers.

Striping

The extraction of fibers from the bark after the retting process is called stripping. The retting time depends upon the harvested crop and the diameter of the steam which is grown. In retting after certain amount of time, it has been verified whether the fibers can be extracted from the bark of the jute plant.

The fiber stripping can be done in two ways

1. Single reed Method
2. Break-Break-Jerk Method

Single Reed Method

In this method, the stripper will take a single or group of (collectively 4 to 5) stems and strip the fiber from the jute steam after the retting process. In this, single long continuous fiber strands are extracted. Fiber extraction takes a long time and is a more tedious process.

Break-Break-Jerk Method

In this method, the stems plant are cut or broken into definite lengths after steeping then the stripper will extract the jute fiber from the stems. The fiber length is small compared to the single reed process.

Washing and Drying

Washing removes dirt, dust, and gummy substance on the jute stem and is an important process to be carried out in jute fiber extraction. Jute has a considerable amount of dust, dirt, and gummy substance as it is undergone in retting. The gummy substance will lie on the material and has to be entirely removed from the material, so thorough washing is required. In a few cases, the jute steam is treated with tamarind water for 10 to 30 minutes and washed in clean water. After squeezing, excess moisture in the fiber is removed by hanging it on a bamboo raft and then taken for the drying process. The process for treating the jute with tamarind water is to remove the dark colour of the fiber. This is being carried as and when according to the quality of the fiber.

The fibers are not to be dried on the ground and spread on a bamboo raft, as wet fibers can easily catch dirt and dust on the ground. These dried fibers are sent to the bundling process.

Bailing and Packing

Jute fibers are graded according to the qualities of the jute extracted; these katcha bales weigh from 220-250 pounds, normally used in markets. The katcha bales are transported to the market and spinning mills.

Jute Fiber Spinning

Spinning is the process of converting the fibres into yarn. The spinning process involves cleaning fibres, opening of fibres, individulising and finally drafting and twisting into the yarn. Compared to the spinning of the cotton fibre the spinning of the jute fibre is different. The jute yarn is spun either from the sliver or from the roving for fine quality of jute yarn. As cotton is soft and more flexible compared to jute fibre, there is a small difference in manufacturing machines' designs and the processing. The process flow of the jute fibre manufacturing is shown in Figure 2.

Batching

Batching is the process of mixing fibers of the different grades received from different sources. Natural textile fibres, including the jute, vary in length, colour and fineness even from the same plant source. The variation is furthermore among the fibres obtained from the different fields with variations in the climatic condition, soil composition, light intensity, and irrigation techniques. To get uniformity in the yarn, and for the optimum utilization of different qualities of fiber the batching is done. In the batching process, keeping in mind the optimization of quality and cost, the fibers of different grades must be mixed in appropriate proportion.

The spinning mills receive the jute fibres in the form of bales from different locations. The first step in initiating the batching is to grade the fibers received from the various locations. Generally, the grading of the jute fibre is done by a skilled and experienced person by subjective assessment. The colour and defects are assessed by visual examination, and the strength, fineness, and length are by handling the fiber tuft. As per the Jute Corporation of India, the jute fibre are mainly classified into five different categories from TDN_1 to TDN_5. The TDN_1 fibres are of superior quality fibres to be used for making the yarn of good quality, while TDN_5 are of inferior quality fibre which may not be suitable for the yarn formation. The different fiber grades are numbered and kept with stock entry at the storage. Based on stock availability and the demand for yarn quality using statistical approaches like linear programming, the fibre stocks are used effectively.

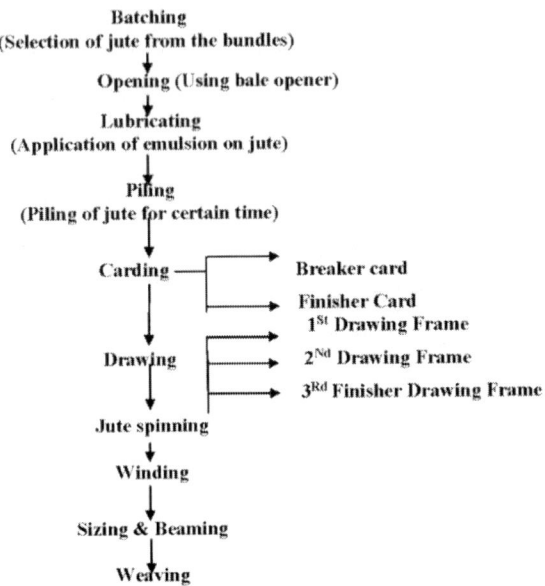

Figure 2. Process flow for the jute yarn manufacturing to weaving.

Opening and Cleaning

Yarn formation requires to get open and clean fibre which are suitable for further processing. After selecting the jute bales to be used for the particular yarn quality, the fibres from the bales are opened. The bale opener is used to open the fibres from the bales stage to the fibre tuft and flees. Thus, the solid slabs of jute fibre are opened and made into a subsequent pliable stage. In this process of opening, dirt, trash, sand and other foreign material are removed and the fiber bulk is cleaned. Different types of bale opener works on mechanical beating and other principles are available. These bale openers are structured with heavy fluted or knob-like projection rollers. As the jute fibre can withstand the tough mechanical action, the jute bale openers may be designed to work even with extra weights too. The surface speed of the rollers is around 30ft/min, and the production is about 60 kg/min. The fluted rollers in the machine revolve slightly slower than the fluted roller at the feed end. The reason behind this is to have a "Dwell" period for the material while in opening, and the material should break/open thoroughly. The modern opener is equipped with a sensor and electronic weighing system to get accurate quantitative weighing and feeding.

Lubricating

To spin the fibre efficiently, the flexibility of the fibre plays essential role. The soft and flexible fibre, like cotton are flow smoothly and processed efficiently without adding any external finish. But for the stiff fibre like jute for efficient processing, it is required to make it soft and smooth. To make the jute fibre soft and smooth, the fibre is treated with the softening agent and lubricated. Lubricating is the process of applying emulsion to the jute fibers for smooth processing. The smooth processing will result into efficient drafting, less fibre breakage, and moist fibre with electric charge dissipation. Two liquids are mixed in such a way that one is dispersed in small round particles in other substance which is assisted by a chemical agent called emulsion. Different emulsions are made using different oil and from the different sources used for lubricating. The emulsion of type oil in water (o//w) is preferred for jute lubrication, and the oil used for the lubrication is generally referred to as JBO (Jute batching oil). Technologists and scholars have been exploring the application of oils, and emulsifying agents to prepare a perfect method of preparation of emulsion on jute fibers like Rice Bran Oil (RBO) and glycerin [3], eco-friendly oil [4].

The emulsion of type oil in water is prepared by adding 20-25% of the oil to 75-80% of water using around 0.5% emulsifier. The oil used for the preparation should be odorless and should have a low degree of saturation with the required viscosity. Also, the oil should withstand the heating condition and should not be affected by climatic conditions. The bulk-level emulsion is prepared in the industry using an open tank with a mechanical stirrer arrangement. A suitable speed is to be set; if not, the emulsion will break with very low or very high speeds. The stability of the emulsion is also a very important factor that is up to what time the emulsion stands on the material without any breakage. The emulsion breaks when the separation of free oil takes place. A broken emulsion cannot serve its purpose. To get a stable emulsion, it is essential to build up the emulsion slowly by successive additions and mixings from time to time.

The emulsion should be applied in the required concentration to the fibre. The application of emulsion may also affect the fabric strength, finishing properties and dyeing process. Generally, the emulsion is applied to the jute stack by drip feed system, saturation, or by spraying through the nozzles.

Piling

After applying the softener, it is required to keep the treated fibre in ambient condition for a sufficient time to get the softening effect on to the fibre. The treated jute fibres are taken away from the softener machine and piled up at a storage place. Usually time given for the piling process is around 2 to 3 days. The jute fibers are wrapped in polyethylene coverings in this process to provide an isolated thermophilic atmosphere [5]. During piling, superficial moisture penetrates inside the fiber and the hard lignin portion gets flexible, and soft. After the piling the fibres are passed through the pile breakers to make it ready for further processes like carding. After the pile-breaking operation, the root cutting is done near the hard feed breaker of the carding machine. The root varies from 5 to 7 percent of the weight of the jute.

Fiber Blending Machine

Sometimes the jute fibre is mixed with other textile fibres to get the hybrid jute yarn with the optimum properties. The fibres such as flax, banana, silk, and cotton can be blended with the jute [6]. Even for some instances, the short waste fibre is mixed to recycle—fiber blending machine designed for homogenous mixing, processing and handling of staple fibers at the carding process. The fiber blending machine is simple in design and fitted to any cards. This machine is available in both hand feed and automatic control feed models. The machine with hand feeding is suitable for sacking weft section using mixed batch of caddies, thread waste and cuttings, etc. This machine helps reduce short fibers' droppings to some extent.

Carding

The next level of fiber opening is carding, where the fibres are separated and individualized. Carding process is called the heart of spinning and the quality of yarn depends greatly on the sliver quality. In this carding process, the jute fibers are passed through high speed pinned rollers moving opposite direction. The carding action is carried by a carding machine's stripper and cylinder rollers. During the carding, the fibre fibers are broken down into an individual fibre and delivered in web form, which is subsequently drawn off through trumpet in the form of ribbon. Also during the carding, microlevel of cleaning and removing entangled mass also took place. The carding is carried out in two types: carding machines: breaker card and finisher card. At the first stage of breaker card the long reeds of jute fiber are broken down with carding action. Twelve slivers coming from the breaker cards are then fed into the finisher card, which is similar to the breaker but has finer teeth and is drawn out into a finer smoother strand.

Drawing

After the carding process, the jute sliver is taken into the drawing frames. Where the slivers are attenuated by means of pairs of rollers to as well as doubled with other slivers. The drawing leads an increase in fibre orientation and uniformity. Four to six slivers from the finisher cards are put through the first drawing frame and converted in one small sliver. Two to six slivers are then put through a second and finer drawing frame, and further combed and drawn out into one sliver. The cans of slivers are then taken to the roving frames to continue the manufacturing process. The draw frames which are used are 1st drawing frame 2nd drawing frame, and third drawing frame. 1st and 2nd drawing frame machines are used for a coarser count of yarns, and all three frames together are used for producing finer or higher counts of jute yarns.

Roving Frame

This is a machine there in sequence after the finisher drawing frame, and this has been used in the olden days of jute manufacturing in modern days, it is being eliminated in the jute manufacturing process. The slivers from the finisher draw frame are converted into "rove" for further spinning. Two to four slivers of the finisher draw frame are taken into the machine and converted into a form of rove, which is used for the further subsequent process of spinning. This process is eliminated in modern spinning mills, which add the further operation after the finisher draw frame. The spinning carried by the rove sliver produces finer and superior quality jute can yarn can be produced. In this spinning process, after passing through three drawing frame, slivers are converted into roving by a roving frame, and after that, a spinning frame is used to produce yarn.

Spinning Frame

The spinning frame is the last machine where the fibre assembly either in the form of roving or sliver is converted into the yarn. During the spinning of the staple fibre at the spinning frame the operations of drawing, twisting, and the winding is carried out. The fibre assembly is further attenuated and to impart strength the twisting is done and finally the yarn is wounded on to the bobbin. Mainly the modified ring frame is used for spinning the jute fibre, and now a days, even rotor spinning is used [7].

The introduction of modern machines has improved the spinning yarn characteristics and properties. Different types of controlled draft spinning machines are used these draft machines are accompanied by rubber, plastic, fluted and endless apron rollers are used in the conversion of sliver into the continuous yarn spinning process. For jute spinning, two types of spinning process are followed; roving spinning and sliver spinning. In the case of roving spinning, the roving from the speed frame is converted into the yarn. The good quality of yarn in terms of strength, uniformity and linear density is obtained from the roving spun yarn but with less production. While in the case of sliver spinning, the throughput rate is higher, but the yarn quality is comparatively inferior.

Winding

The following process after spinning is the winding process. In spinning, the sliver received from the drafting frames are converted into the continuous yarn, wound on the packages like spools and cones. Winding is the operation of transferring of yarn from one package into another, like spools for warp and cops for the weft. Winding of yarn is necessary for various reasons and purposes. For subsequent beaming operation, the yarn wound on spools for better handling, less breakages, and piecing up less wastage and ease in transportation. The yarns used in weft weaving are wound on cops or pirns. The yarn is also reeled into suitable hanks for dyeing. The yarn is wound on suitable packages like cones, spools, and cheese for further utilization.

Weaving

Weaving is the process of making the fabric by interlacing the warp and weft yarn. The yarns running vertically and parallel to the selvage is the warp yarn and the horizontal is the weft. To weave a fabric it is required to prepare the warp and weft yarn. The series of warp yarns are presented from the beam form with multiple ends as per the requirement of the fabric width and the thread density. The weft yarn is inserted as a single end for each weaving cycle and drawn from the pirn or cone package form. From the winding process, the jute yarn is delivered as a single-end cone package. The first step in the weaving is to make the warp beam by winding the yarns from the different numbers of cones. Weft preparation for the shuttle loom is to wind the pirns from the cone. Before going into the weaving process, the jute yarn is coated with suitable material to impart the elasticity, strength and to bind the hair in it. The tamarind kernel powder (TKP) is mainly used for sizing the jute yarn.

The loom used for weaving the jute fabric is chosen according to the end applications. The handloom fabric are known for its aesthetic look, excellent texture and ornamentation possibilities. for the. The handloom is used to weave the jute fabric and is used for the aesthetic application, mainly the craft-based applications [8]. The handloom can also be used in some of apparel fabric applications. The fabric woven for apparel applications is not purly jute but in the form of blended jute or union fabric [9]. The power loom is known to produce the cloth with more production rate. For the making in more quantity of medium to lower grade like sack bag, the power looms can be employed.

The modern weaving machines are shuttleless looms. The shuttleless looms can be used to produce the fabric of good quality jute fabric. The rapier loom are used for the production of sack cloth and the hessian cloth of good quality. The airjet loom can also produce the jute fabric with a higher production rate. The wide-width fabric used for the technical applications like geotextile can be produced using the projectile loom.

Product and Applications

As discussed the in earlier section the yarn manufacturing and the fabric manufacturing from the jute fibre. The yarn spun from the jute fibre can be used in different forms. The spun single yarn are used for the weaving application. The double and twisted yarn are used for the application of twin yarn which can be explored further for the craft applications. Other than normal spinning process, the spinning process can be modified to produce some of the fancy yarn [10].

The fabric produced by weaving are of different qualities manufactured for the different applications. The coarser fabric of low quality grade are used for sack cloth and for carpet. The good quality of fabric produced from the fine grade of jute fiber are used for the application of hessian cloth and upholstery. The jute fabric are used for technical applications like geotextiles are made in longer width and of open structure. Jute fabric can also be used as the reinforcement fabric for plaster and for fabric-reinforced composites. The jute fibre in the form of fibrous web as a nonwoven fabric also finds different applications, from insulating layers to the mulch mat [11].

Conclusion

In most plant-based natural fibre after cultivation, fibres are directly picked from the specific part of the plant and packed into the bales. While for the extraction of the jute fibre the fibres strands are required to be separated by a process called as retting. The utmost care needs to be given from the initial stage of the process to get the excellent quality of product afterward. Accordingly, based on the fiber's desired quality, the particular type of retting and retting parameters are selected. After retting, washing and post-retting processes, the fibres are finally packed into the bales. The jute yarn is stiffer and harder than the other staple textile fibre. The jute's yarn formation process is slightly different compared to the normal cotton yarn manufacturing process. Special care is given to select a particular type of lubricant and its application to the fibre to get efficient yarn formation.

Further, the jute fibres are broadly divided in to shorter length and good quality fibres of longer lengths. Based on the product type, the fiber's specific quality is chosen. Generally, for product like bags, upholstery, and apparel grade fabric, the good quality of fibre is selected. The shorter fibre can be converted into a fibrous sheet of nonwoven fabric and used for the composite or for filling material. Sustainability point of view for the technical application jute is a promising fibre. The natural-based fibre with good mechanical properties and durability.

References

[1] Zakaria Ahmad, and Firoza Akther. (2001). "Jute retting: An Overview," *Online J. Boil. Sci.*, vol. 1, no. 7, pp. 685–688.

[2] Ali, M. R., Kozan, O., Rahman, A., Islam, K., and Hossain, M. I. (2015). "Jute retting process: Present practice and problems in Bangladesh," *Agric. Eng. Int. CIGR J.*, vol. 17, no. 2, pp. 243–247.

[3] Siddiqua, T., Begum, H. A., Siddique, A. B., and Stegmaier, I. T. (2021). "Effect Of Oil And Oil - Free Emulsion On Jute Fiber Processing And Yarn Properties," *Int. J. Sci. Technol. Res.*, vol. 10, no. 04, pp. 237–239.

[4] Pankaj, F., Sivasurain (Inidan Oil Corporation Ltd, Research & Development Centre, "Composition of eco-friendly oil for jute batching application," 2012 [Online]. Available: https://patentimages.storage.googleapis.com/ee/38/c7/318788 941c91ad/WO2012066573A2.pdf.

[5] Shyamal Banik, S. N. G. (2008). "Pectinolytic activity of microorganisms in piling of jute," *Indian J. Fibre Text. Res.*, vol. 33, no. 2, pp. 151–156, [Online]. Available: https://www.researchgate.net/publication/289088581_Pectinolytic_activity_of_mic roorganisms_in_piling_of_jute\.

[6] Basu, G., and Roy, A. N. (2008). "Blending of Jute with Different Natural Fibres," *J. Nat. Fibers*, vol. 4, no. 4, pp. 13–29, Mar., doi: 10.1080/15440470801893323.

[7] Sunil Kumar Sett, D. S. (1993). "Mechanical Behaviour of Rotor Spun Jute - Viscose Blended Yarns at different Twist Levels," *Indian J. Fibre Text. Res.*, vol. 18, no. 0, pp. 20–24.

[8] Alok Nath Roy, G. B. B. (2010). "Improvement of a traditional knowledge by development of jacquard shedding based handloom for weaving ornamental jute fabric," *Indian J. Tradit. Knowl.*, vol. 9, no. 3, pp. 585–590.

[9] Sankar Roy Maulik, "Value added fashion apparels made of jute," *Asian Dye.*, vol. 16, no. 1, pp. 55–58, 2019, [Online]. Available: https://www.researchgate.net/ publication/336179443_Value_added_fashion_apparels_made_of_jute.

[10] Surajit Sengupta, S. D. (2010). "A new approach for jute industry to produce fancy blended yarn for upholstery," *J. Sci. Ind. Res. (India).*, vol. 69, no. 12, pp. 961–965.

[11] Maity, S., Singha, K., Gon, D. P., Paul, P., and Singha, M. (2012). "A Review on Jute Nonwovens: Manufacturing, Properties and Applications," *Int. J. Text. Sci.*, vol. 1, no. 5, pp. 36–43, doi: 10.5923/j.textile.20120105.02.

Chapter 5

Jute: Potential Applications in Projects of Environmental Recovery and Environmental Conservation

L. H. Tsuchiya and A. M. Da Silva
Department of Environmental Engineering – Institute of Sciences and Technology of Sorocaba, Sorocaba, São Paulo, Brazil

Abstract

In this work, we research and present some biological and ecological properties of jute. We also present a summarized set of case studies and we performed a PESTLE analysis to facilitate the presentation of some results. We have found that there is a vast potential for the use of jute-based products, both as fibers and as live plants, featuring an open range of business opportunities and new research. Regarding the potential for new scientific works, we address about the necessity to improve the production method, in order to make the production system truly sustainable. Also, there is a demand to use fibers in their most natural state as possible, because although jute fiber is biodegradable (perhaps its best ecological property), the fact of applying synthetic products to improve the quality of the product can make it as harmful to ecosystems as synthetic fibers.

Keywords: jute fiber, environmental benefits of jute, naturalistic engineering

In: Jute: Cultivation, Properties and Uses
Editor: Matthicu Issa
ISBN: 979-8-88697-490-4
© 2023 Nova Science Publishers, Inc.

Introduction

The environmental crisis that we have been experiencing started more than 100 years ago and it has forcing and stimulating humans to seek new products able of replacing old contaminant products by sustainable and/or products for helping to solve others types of environmental problems. The worldwide disposal of millions of tons of materials such as plastics, rising global temperature and rapid depletion of oil resources, rising sea levels constitute some examples. Such problems constitute arguments for demanding the development of green and sustainable products that are gradually endorsed for sustainable development (Singh et al., 2018).

One category of product that is commonly used for several human aims is fibers. Fibers might be artificial (or synthetic) or natural. Artificial fibers are used for an infinity of purposes and are obtained from the transformation of natural polymers, through the action of chemical agents, in extrusion processes for example. Some kinds of fibers (nylon or polyester, for example) might be used to increment the mechanical resistence of the concrete and do not harm the quality of the material by corrision or similar process. However, they have caused several environmental and human health damages, once they have excessively been deposited in some places, especially oceans as microplastics (Carney Almroth et al., 2018).

One alternative is using natural fibers. Natural fibers are usually of vegetable origin, coming from different parts of plants such as seeds, stems and even flowers and leaves. Less usual, they also might be of animal origin, obtained from or produced by animals. They are usually eco-friendly especially because of the process of decomposition and assimilation by the ecosystems.

Vegetable fibers, such as sisal (*Agave sisalana* Perrine), jute or patsun (*Corchorus capsularis* L.) and of mallow (*Urena lobata* L.), are renewable natural resources, of low cost and do not cause damage to human health such as asbestos fibers. Therefore, with regard to materials used as reinforcement, can be an innovative and effective solution to meet the Sustainable Development Goals – SDGs.

However, natural fibers, although considered environmentally friendly for their faster degradation, can be a global threat comparable to synthetic polymers. In fact, as a result of textile processing, they can be mixed with flame retardants and/or resins, which not only poses a threat related to the release of toxic compounds, but also affects the speed of degradation, which becomes longer (Santini et al., 2022). Hence, the best way is trying to use the material as natural as we can, to avoid the risks of contamination.

We will see ahead that one of the most common materials used as natural fiber is the jute. Hence, in this work we had the goal of survey and presenting the characteristics of the jute and analyzing the potential of the product for environmental purposes. To facilitate the analysis and the presentation of the results, we adopted the PESTLE system of analysis.

The Jute: Biological and Ecological Characterization

It is an C3, annual herbaceous vascular plant, presenting long, soft, shiny vegetable fibres. In terms of Botany, it belongs to the Malvales order / Malvaceae family. The origin is from Asian tropical regions and it is cultivated mainly in Asia and Africa (Saleem et al., 2020), although in South America it is also frequent. In Brazil, the species was introduced by the Japanese community in the Amazon region at the early of the 20th Century.

History shows that the main occupations were the floodplain areas of some rivers, which on the one hand benefited some small rural producers, but on the other hand, caused deforestation in several riparian areas (Homa, 2016). There were also attempts to grow jute in the state of São Paulo (acronym: SP), whose seeds came from Calcutta. Despite numerous attempts, this culture was not successful in SP. Its success actually occurred in the Amazon region specifically. Due to the climate, the juta quickly adapted to lowland agriculture in the Amazon, becoming an integral part of the region's natural resources, particularly in the state of Amazonas. The possibility of it being introduced in the floodplains of the Amazon River was envisaged due to the similarity with which the cultivation of jute was conducted in India, especially on the banks of the Ganges and Brahmaputra rivers (de Oliveira Costa Filho et al., 2022).

Jute fiber is collected from the outer region of the stem. It occurs after the maceration of the entire plant and constitutes an important source of different degrees of pulp (Del Rio et al., 2009). Jute plants produce a very valuable, versatile and useful material known as lignocellulosic fibers. They are multicellular characteristic products. The main constituent components are lignin, cellulose, oils, waxes and some types of fats (Ashraf et al., 2019).

In jute fibers, the amount of lignin, hemicellulose and cellulose directly interfere in the properties. Generally, the higher the amount of lignin, the worse the mechanical and thermal performance. The cellulose fraction is hydrophilic, and does not develop the best fiber-matrix interaction with hydrophobic polymer matrices. Cellulose appears to be the main component. It is found in the form of thin rods aligned along the length of the fiber. It is a polysaccharide, consisting of a semicrystalline linear chain with hundreds of β-(1-4)-glucosidics linked with D-glucopyranose in the presence of hydroxyl groups (OH⁻). In turn, hemicellulose is a lower molecular weight polysaccharide, whose function is to be a matrix for cellulose in plant walls. Although cellulose is crystalline, hemicellulose has a random structure and is therefore amorphous. Lignin is a class of complex polymers of hydrocarbons, which gives rigidity to the plant, being relatively hydrophobic (Ribeiro, 2018).

The Potentials of the Jute as an Environmentally Friendly Material

In the past, high manufacturing costs and lack of synthesis methods usually restricted the growth of biocomposites. However, environmental concerns have increased in importance. The remarkable potential of jute-formed composites to provide environmentally friendly materials is the essential driving force behind their rapid development (Figure 1). Jute-based composites have attracted the attention of researchers, as well as attracting research and development funders, due to their best physical and mechanical properties among all-natural fibers (Ashraf et al., 2019).

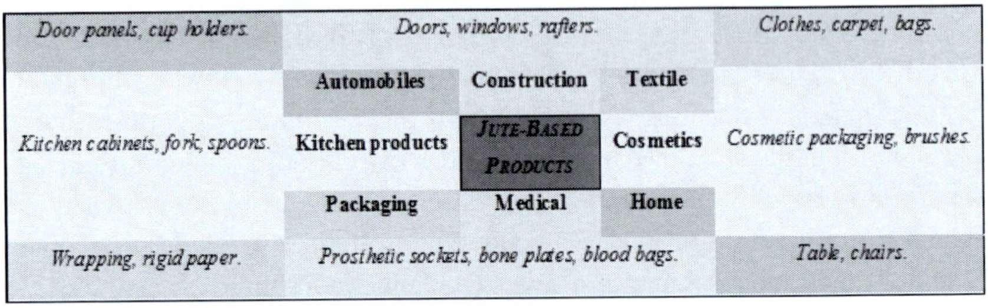

Door panels, cup holders		*Doors, windows, rafters.*		*Clothes, carpet, bags.*	
	Automobiles	**Construction**	**Textile**		
Kitchen cabinets, fork, spoons.	**Kitchen products**	***Jute-Based Products***	**Cosmetics**	*Cosmetic packaging, brushes.*	
	Packaging	**Medical**	**Home**		
Wrapping, rigid paper.		*Prosthetic sockets, bone plates, blood bags.*		*Table, chairs.*	

Source: Modified from Ashraf et al., (2019).

Figure 1. Some examples of jute-based products. Bolded terms: areas of applications. Italic terms: products related to the areas (following the colors of the areas).

In Brazil, as in other regions a significant part of the jute tissue is detined to fabrication of bags of coffee, potatoes and in a lesser amount, bags of peanuts, cocoa, nuts, and tobacco. In the market of handicraft, the jute is commonly and extensively used to diverse ends. Furthermore, jute fibres have replaced synthetic fibre in the composition of homes such as carpets, ropes and sacks (Saleem et al., 2020). A small portion of the total production is effectively destined to industrial or environmental uses, but the product has attracted the marked especially due to the potential of biodegrability.

Potential Use of Jute Especially as Part of Solution for Environmental Problems

Study Cases

As mentioned earlier as an alternative for several industrial ends and also for projects of environmental recovery or conservation. We selected and introduced ahead some case studies aiming to illustrate the versatility of jute. As exemplified, we see that jute has the potential to be used while a living organism (not only fibers) to solve problems of environmental contamination.

Case 1

Comparison of the performance between fiberglass and jute fiber in a physical system of ceramic cladding on ventilated facade (Ribeiro, 2018). The author studied the application of a jute fiber composite in a physical system of ventilated facade compared to the current system of fiberglass, in terms of characterization of the constituents, mechanical characterization of the systems, the mechanical performance after degradation in water and the behavior of the composite system when exposed to a heat source. According to the author, regarding the physical characterization of the composite system and its constituents, the trials that showed important information were: thermogravimetric analysis, the tensile strength of the fiber without the ceramic and water absorption of the fiber. The thermogravimetric analysis performed by Ribeiro (2018) showed that the residual mass of the epoxy resin refers to the mineral filler, while the fiberglass mass loss is related to the polymeric layer. Already the Jute fiber has the highest mass loss due to fiber origin. The fiber tensile tests indicated that the glass fiber has highest breaking load in the 90° orientation and the highest resistance to traction at 0°. Jute fiber has a tensile strength much lower than fiberglass (88% lower), however, the behavior during test, shows that the jute fiber presents the rupture of the fibers of gradually. Regarding water absorption, jute fiber has absorption 6 times greater in relation to fiberglass, being able to compromise the mechanical performance of the fiber when inserted into a polymer matrix. The mechanical characterization tests of the configurations indicated that the measured fiber-matrix adhesion is greater than the fiber tensile strength regardless of orientation. Already the tests of tensile and flexural strength of the composite set, were not relevant for the evaluation of the resin or fiber, being more appropriate the composite tensile strength test. The author states that, although the jute fiber presents inferior mechanical performance, the composite set with jute fiber presents the minimum normative requirement. However, the flame propagation, degradation and water absorption performance tests of the composite show that the jute fiber cannot be applied in a physical ventilated facade system, as the jute fiber makes the set vulnerable. Regarding the cost of fiber, both have similar values in the market (national natural fiber and imported glass fiber). However, comparing the values between jute fiber and national glass, there is a 5% reduction with the use of natural jute fiber in the composite.

Case II
Fiber length and relationship with the acoustic behavior of jute fiber (Sambandamoorthy et al., 2021). The authors studied the influence of fiber length and surface alteration by chemical treatment on the acoustic characteristics of jute fibers. The study attests that jute fiber has excellent sound absorption capacity and therefore can be used for noise reduction and control application without affecting the environment. The sound absorption coefficient increases as the fiber length increases. The authors found that the untreated jute fibers showed better sound absorption capacity than the treated ones, since they argue that the treatment reduces the structure of the hollow lumen as seen in the SEM images.

Case III
Use of jute as a conditioner or recuperator of degraded soils – Here we illustrate this quality of the jute presenting three related study cases:

(a) First, in a study conducted by Tarafdar and De (2018), the authors created pieces of jute that they called "agrotextiles" and analyzed the potential for maintaining soil quality and protecting the soil against any form of degradation. They argue that jute agrotextiles improve soil properties and attribute increased yields to crops such as tomatoes. Improvements in soil bulk density, porosity, moisture, as well as better aggregation and stabilization of soil aggregates occurred due to the application of each strength of the jute agrotextiles. They suggest that 600 g.m^{-2} of jute agrotextiles could provide the necessary effectiveness to guarantee beneficial effects to tomato cultivation and reflect in greater yield and cost-benefit optimization, in addition to improving the ecological conditions of the soil used for cultivation, ensuring its continuous conditioning or its conservation.

(b) In another project Tsuchiya (2021), studied the performance of the jute as a part of a product named "bioblanket" for simultaneously control the soil erosion and control the growing of weeds in cleaned lots. The study was conducted in Sorocaba, São Paulo, Brazil, on a high slope slope (59°), delineated by an experimental area composed of 3 plots with jute bioblanket and 3 plots in control condition (Figure 2). Normally, engineering works contemplate the use of a single layer of jute fabric placed on soil surface. Here the author filled a double layer of jute fabric with dry grass (Figure 3). The results obtained by the analyzes demonstrated the efficiency of jute bioblanket in reducing 30% of the growth of invasive vegetation, and in protecting soil against nutrient leaching. The application of geotextile also helped to retain water, maintaining the condition of soil moisture. Jute fiber's characteristic to absorb water, makes the material interesting for engineering projects, as it reduces the impact of rain on the ground. It is worth mentioning that jute fiber is a material found more easily in countries such as Brazil, India, and Africa. Jute fiber is extracted from the *C. capsularis* plant, using a process of drying and braiding the filaments using wax.

Source: Tsuchiya (2021).

Figure 2. Experimental plots using jute bioblankets for erosion control and also control the growing of weeds.

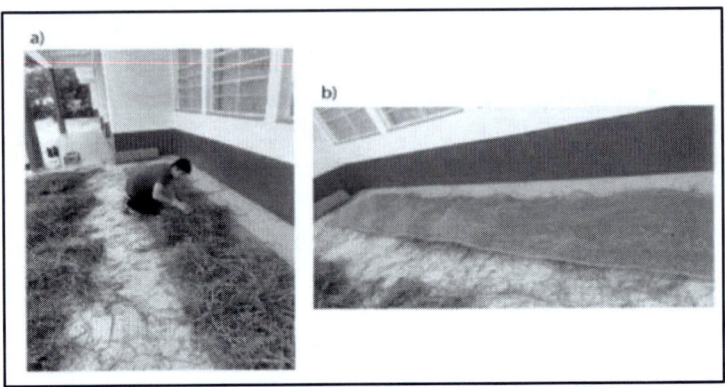

Source: Tsuchiya (2021).

Figure 3. Steps of manufacturing the bioblanket using jute tissues and dries straw.

(c) The third study-case for this issue is concerning the studies conducted by Álvarez-Mozos et al., (2014a and 2014b), who monitored the rates of soil loss and growth of vegetation in experimental plots. In terms of soil control, the authors observed that the jute and buried geogrid produced lower erosion rates (average soil loss of 3.2 g m^{-2} and 2.1 g m^{-2}, respectively) compared to control plots (average soil loss of 3.6 g m^{-2}) at 45°. However, at 60° the observed erosion rates were similar to the control. In terms of plant growth, the authors found initial establishment was 2 to 3 weeks faster for the geogrid treatments than for the control, both in plots installed at 45° and in those installed at 60°. They noted contradictory results in the treatment conducted with the mesh made with jute in the establishment phase for both plots with an increased coverage on the slope of 45°, but a decreased coverage on the slope of 60°, when compared to the control plots. The blanket made with coconut material significantly impeded the growth of vegetation on both slopes. For the remaining period of the experiment, the authors did not observe significant differences between the geogrid and the control for the plots at 45°. However, geotextiles manufactured with biological materials resulted in lower plant coverage when compared to the control and also to jute and coconut. The authors argue that reduced vegetation growth in the jute and coconut plots was due to the growth of runoff in the steep plots. In the case of coconut blanket, the reduced growth was also due to the high percentage of coverage of the material that blocked the contact between the plants and the soil. They concluded that polyester (artificial fiber) geogrids are recommended for a joint beneficial effect on erosion control and vegetation growth on steep slopes hydroseed with compacted soils in areas with a similar climate.

Comparing the results of studies carried out by Tarafdar and De (2018), Tsuchiya (2021), Álvarez-Mozos et al. (2014a) and Álvarez-Mozos et al. (2014b) we can observe as a trend that jute, when used as a soil protector and conditioner, it has the potential to act positively in the soil. Regarding the growth of vegetation, the results are very interesting and they are heading towards a same common point, because it seems that jute is an effective inhibitor of vegetation. So, for works where the objective is to limit the growth of vegetation, it has full potential, while for cases where there is a desire to facilitate the growth of vegetation, it may be better to use other alternatives. Another point highlighted in the works is about the time span of durability of the product. As it is biodegradable, the shelf life is not very long (much shorter than synthetic fibers). This we interpret as a quality, rather than a limitation. Hence, in cases where the soil is still poor in vegetation cover, there would be a necessity to replace a new covering of jute-based products as that one proposed by Tsuchiya (2021).

Case IV

Use of jute as an alternative for phytoremediation projects. Phytoremediation constitutes an option of environmental cleaning through the use of plants and their related microorganisms (Pilon-Smits, 2005). In this case, the jute plants are living organisms and the physiological processes that occur inside the plant and ecological interactions that occur in the region of the root system and contaminants are the main focus of interest. Exemplifying, jute plants can jute can absorb different heavy metals from every source, i.e., soil, water, and a field environment. Plants are able to survive in environments with high concentrations of cooper in the soil, because although they may suffer oxidative damage to the leaves of jute plants, the plant organism is able to overcome this adversity through the action of antioxidant enzymes. The plants also have ability of extract heavy metals from sediments, suggesting to have some potential for sewage treatments (Niazy et al., 2017; Sallem et al., 2020).

The Potential of Use of Jute Assessed by the PESTLE

A Glance about the PESTLE Analysis

Before we proceed with the analysis, we understand that is important a brief explanation about the PESTLE.

Initially, PESTLE is an acronym for political (P), economic (E), social (S), technological (T), legal (L), and environmental (E). Therefore, PESTLE analysis is used to analyze political, economic, sociocultural and technological changes in the business environment. It helps to bring a more macro view of the external threats and opportunities existing (Perera, 2017). A brief explanation of each factor is exposed ahead (Perera, 2017; Ulubeyli and Kazanci 2018):

Political factors - Political factors are important to be analyzed, because in addition to knowing political stability or instability, it also talks about government policy and foreign trade, corruption, labor, environmental legislation and trade restrictions, taking into account the level of government interference in the economy. Hence, it permits to show how attractive the market is and can even show the government's impacts on health regulations, infrastructure and education system, for instance.

Economic factors – by this factor we might to focus on analyzing economic growth, inflation rates, interest rates, exchange rates and even unemployment rates.

Social factors – it helps to analyze the cultural influences and beliefs of the target audience to better understand the social environment in which the company is operating. Demographics, norms, customs, age, interests, societal opinions and education are also analyzed, these factors are especially important for marketers to segment their customers.

Technological factors – they are usually one of the most important, as they determine how companies can explore communication and interaction, define the level of innovation activity and technological knowledge that the market has.

Ecological (or Environmental) factors - currently consumers are more concerned about the environment, so they choose to buy products from socially responsible companies with ecology. Therefore, analyzing these factors is important for an organization. Ecological factors include environmental aspects such as the climate and its environmental changes and tradeoffs. Thus, many companies pursue sustainability and corporate social responsibility.

Legal factors - These factors relate to product safety issues, equal opportunity law, consumer rights, labor laws, copyright laws, discrimination laws, health and safety laws. Legal factors serve for companies to function successfully, knowing the laws and regulations of the country and respecting them.

The PEST methodology helps uo to: Detect threats; Detect opportunities; Develop an objective view of an organization's external environment. The tool can be used when planning the next steps towards the analysis of a product (or even a company), to better understand the environment that is inserted and the possible impacts of the analyzed factors or when starting a new business.

Analysis
We did a PESTLE analysis especially considering the Brazilian situation and considering the manufacturing and application use of the bioblanket. We verified that there is a clear potential for the use of jute-based products, especially the bioblanket in places where there is a need to mitigate soil degradation problems, both in rural and urban areas (Figure 4). This characterizes an open range of business opportunities.

Political	Economic	Social	Technology	Legal	Environment
Increase environmental policies with the application of natural technologies.	Brazil is an important producer of jute, especially because the tropical climate conditions; Decreased labor costs due to biodegradability factors. Necessity of economical incentives for producting the commodity.	The jute bioblanket has the potential to minimize the effects of soil degradation that can affect society; Its use can be replicated for projects of environmental recovery and protection of agricultural crops. Potential use in agroforestry systems.	Jute plants requer small quatities of fertilizers for cultivation. The bioblanket filled with pruning residues increases the potential of the jute fabric, in addition to being easily transported and positioned on high slope terrain.	Brazilian Association of Technical Standards (NBR); Brazilian Federal Laws concerning protection and conservation of the Natural Resources.	Jute is a biodegradable material; Jute protects the soil against erosive processes; The bioblanket is able to control the growth of invasive vegetation.

Figure 4. PESTLE analysis about the jute-based bioblanket.

The analysis revealed that in Brazil there is no legislation regarding jute, nor technical regulations that specifically guide or even specific economic incentives for the use of products based on jute fibers. Complementary qualities that highlight the potential use of jute are that growing jute requires low amounts of fertilizer (Islam & Ahmed, 2012). There is also the potential for the use of jute leaves as food, since they contain compounds that produce numerous medical and health benefits that promote the general well-being of the body (Ahmed & Sarkar, 2022), although in Brazil there is no use for this end (eating).

Final Remarks

Jute is a plant species with notorious importance for humans and has been used for decades. Even so, it has great potential for use for purposes other than those already established. If it has wide and great utility, then it has wide production demand.

Therefore, research could be continued with two major goals:

(i) to prospect sustainable forms of production - for example, considering the use of jute in agroforestry systems (Jewel and Rahman, 2015) can be an alternative to improve land cover and provide extra income to the producers of this commodity;
(ii) cellulosic fibers should be used as pure as possible. Otherwise, even if they do not constitute an environmental problem in themselves, any additives or dyes within them can be potentially carcinogenic and harmful to organisms in different ecosystems and, consequently, to humans (Santini et al., 2022).

Jute has strong potential to be used in a "pure" state for various purposes.

References

Ahmed, Z., & Sarkar, S. (2022). Review on jute leaf: A powerful biological tool. *International Journal of Scientific Research Updates*, 4(1), 64 – 85.

Álvarez-Mozos, J., Abad, E., Giménez, R., Campo, M. A., Goñi, M., Arive, M., Casalí J, Díez J. & Diego, I. (2014a). Evaluation of erosion control geotextiles on steep slopes. Part 1: Effects on runoff and soil loss. *Catena*, 118, 168-178.

Álvarez-Mozos, J., Abad, E., Goñi, M., Giménez, R., Campo, M. A., Díez, J., Arive, M. & Diego, I. (2014b). Evaluation of erosion control geotextiles on steep slopes. Part 2: Influence on the establishment and growth of vegetation. *Catena*, 121, 195-203.

Ashraf, M. A., Zwawi, M., Taqi Mehran, M., Kanthasamy, R., & Bahadar, A. 2019. Jute based bio and hybrid composites and their applications. *Fibers*, 7(9), 77.

Carney Almroth, B. M., Åström, L., Roslund, S., Petersson, H., Johansson, M., & Persson, N. K. (2018). Quantifying shedding of synthetic fibers from textiles; a source of microplastics released into the environment. *Environmental Science and Pollution Research*, 25(2), 1191-1199.

de Oliveira Costa Filho, A., Lasmar, D. J., da Silva Chaar, J., Oliveira, R. F. P., & Lima, J. A. Q. (2022). Ouro da várzea amazônica: panorama e estímulo para o cultivo da fibra vegetal de juta (*Corchorus capsularis*) e geração de emprego para os ribeirinhos no estado do Amazonas [Gold from the Amazon floodplain: panorama and stimulus for the cultivation of jute vegetable fiber (*Corchorus capsularis*) and job creation for riverside dwellers in the state of Amazonas]. *Brazilian Journal of Development*, 8(4), 31423-31438.

Del Rio, J. C., Rencoret, J., Marques, G., Li, J., Gellerstedt, G., Jimenez-Barbero, J., Martinez, A. T. & Gutiérrez, A. N. A. (2009). Structural characterization of the lignin from jute (*Corchorus capsularis*) fibers. *Journal of Agricultural and Food Chemistry*, 57(21), 10271-10281.

Homma, A. K. O. (2016). *A imigração japonesa na Amazônia: sua contribuição ao desenvolvimento agrícola* [*Japanese immigration in the Amazon: its contribution to agricultural development*]. Brasília, DF: Embrapa.

Islam, M. S., & Ahmed, S. K. (2012). The impacts of jute on environment: An analytical review of Bangladesh. *Journal of Environmental Earth Sciences*, 5, 24-31.

Jewel, K. N. E. A., & Rahman, M. M. (2015). Late Jute seed production in cropland agroforestry system. *Azarian Journal of Agriculture*, 2 (6), 162-166.

Niazy, M. M., & Wahdan, M. E. M. (2017). Enhancing phytoremediation of Pb by treating soil with citric acid and growing white jute (*Corchorus capsularis* L.), and river red gum (*Eucalyptus camaldulensis*). *Zagazig Journal of Agricultural Research*, 44(4), 1359-1367.

Perera, R. (2017). *The PESTLE analysis*. Nerdynaut.

Pilon-Smits, E. (2005). Phytoremediation. *Annual Review of Plant Biology*, 56, 15.

Ribeiro, A. (2018). *Caracterização e aplicação de compósito de fibra de juta em sistema de revestimento cerâmico em fachada ventilada* [*Characterization and application of jute fiber composite in ceramic coating system on ventilated façade*]. MSc Dissertation. Universidade do Sudoeste Catarinense. Santa Catarina, Brazil.

Saleem, M. H., Ali, S., Rehman, M., Hasanuzzaman, M., Rizwan, M., Irshad, S., Shafiq, F., Iqbal, M., Alharbi, B. M., Alnusaire, T., & Qari, S. H. (2020). Jute: A potential candidate for phytoremediation of metals - A review. *Plants*, 9(2), 258.

Sambandamoorthy, S., Narayanan, V., Chinnapandi, L. B. M., & Aziz, A. (2021). Impact of fiber length and surface modification on the acoustic behaviour of jute fiber. *Applied Acoustics*, 173, 107677.

Santini, S., De Beni, E., Martellini, T., Sarti, C., Randazzo, D., Ciraolo, R., Scopetani C. & Cincinelli, A. (2022). Occurrence of natural and synthetic micro-fibers in the mediterranean sea: a review. *Toxics*, 10(7), 391.

Singh, H., Singh, J. I. P., Singh, S., Dhawan, V., & Tiwari, S. K. (2018). A brief review of jute fibre and its composites. *Materials Today: Proceedings*, 5(14), 28427-28437.

Tarafdar, P. K., & De, S. K. (2018). Efficient use of jute agro textiles as soil conditioner to increase tomato productivity. *Journal of Crop and Weed*, 14(1), 122-125.

Tsuchiya, L. H. (2021). *Uso de Biomanta no controle de espécies de plantas invasoras e manutenção da água no solo* [*Use of Biomat in the control of invasive plant species and maintenance of water in the soil*]. MSc Dissertation in Civil and Environmental Engineering. Sao Paulo State Unuversity (UNESP), Sorocaba, SP, Brazil.

Ulubeyli, S., Kazanci, O. (2018). Holistic sustainability assessment of green building industry in Turkey. *Journal of Cleaner Production*, 202, 197-212.

Index

A

adsorbent, vii, viii, 51, 52, 53, 54, 57, 58, 63, 66, 67, 69, 70, 72, 74, 75, 76, 77, 78, 79, 80, 81, 82, 86, 93, 94, 97, 98, 99, 100, 101, 102, 103, 104, 105, 106, 107, 108, 109, 110, 111
antimicrobial properties, ix, 113
automotive, ix, 15, 18, 19, 27, 39, 41, 45, 47, 48, 113

B

bast fibers, ix, 10, 98, 100, 114
batching, 119, 121, 127
binding, viii, 29, 51, 53, 54, 61, 62, 65, 71, 77, 88
biodegradable, vii, ix, 2, 3, 11, 15, 20, 21, 22, 27, 36, 44, 51, 54, 78, 100, 114, 129, 137
biological degradation, viii, 97, 99
biomaterials, viii, 45, 84, 89, 97, 98
break-break-jerk, 117, 118
business, ix, 83, 88, 129, 138, 139

C

carbon, viii, 11, 16, 29, 34, 43, 46, 48, 67, 79, 81, 82, 83, 97, 104, 106, 107, 111
carding, 3, 122, 123
cellulose, viii, ix, 2, 9, 10, 37, 38, 44, 51, 53, 54, 55, 62, 63, 67, 78, 80, 81, 82, 83, 84, 85, 87, 88, 89, 90, 91, 92, 93, 95, 98, 100, 102, 107, 131

chemical, vii, 4, 8, 10, 11, 52, 55, 62, 63, 71, 74, 80, 84, 87, 90, 94, 101, 103, 105, 107
chemical modifications, vii, 4, 8, 10, 11, 52, 55, 62, 63, 71, 74, 80, 84, 87, 90, 94, 101, 103, 105, 107
circular economy package, viii, 52, 76, 78
climate, 3, 114, 131, 136, 138
coagulation, viii, 97, 99
composite materials, vii, 1, 2, 8, 10, 14, 16, 17, 21, 28, 29, 31, 33
compression, 11, 12, 16, 17, 18, 19, 20, 21, 24, 26, 27, 29, 30, 41, 42, 43, 46, 90
construction, ix, 13, 27, 48, 76, 113
Corchorus capsularis L., vii, 2, 51, 53, 78, 100, 130, 140, 141
Corchorus olitorius L., vii, ix, 2, 51, 53, 100, 111, 113
cotton, vii, 2, 51, 53, 85, 87, 90, 91, 92, 93, 94, 95, 100, 119, 121, 122, 126
cross-linking, vii, 1, 10
cultivation, vii, ix, 3, 104, 107, 113, 114, 126, 131, 135, 141
curing, vii, 1, 10, 11, 14, 18, 19, 25, 66, 78

D

desorption, 56, 64, 65, 71, 74, 75, 101, 103, 105, 106, 107, 109
drawing, 123, 124
drylands, 114
dyes, viii, 36, 43, 51, 52, 54, 55, 67, 69, 70, 71, 72, 73, 74, 75, 77, 78, 79, 81, 82, 84, 85, 88, 91, 92, 97, 100, 101, 104, 105, 106, 109, 110, 111, 127, 140

E

economy, viii, 2, 36, 52, 54, 76, 77, 78, 81, 87, 95, 100, 138
environment, viii, ix, 2, 16, 17, 37, 41, 52, 54, 76, 78, 98, 102, 106, 108, 134, 137, 138, 139, 140, 141
environmental benefits, 129
environmental conservation, vii
environmental recovery, vii, 133
environmentally friendly, viii, 22, 26, 51, 54, 77, 79, 82, 114, 121, 127, 130, 132

F

fabric, vii, ix, 1, 4, 5, 11, 12, 13, 17, 18, 24, 25, 29, 33, 35, 37, 40, 41, 42, 46, 47, 48, 55, 56, 58, 60, 61, 62, 63, 65, 66, 68, 69, 71, 72, 73, 75, 76, 77, 78, 80, 84, 85, 86, 87, 88, 89, 90, 91, 92, 93, 94, 95, 113, 121, 125, 126, 127, 135
fats, viii, 51, 53, 114, 131
fiber, vii, 1, 5
fiber blending machine, 122
fiber rovings, vii, 1, 5
fiber-matrix compatibility, vii, 1, 8
fibre reinforced composites, ix, 113
flammable, vii, 1, 8

G

geotechnical, ix, 113
green composites, 1, 17, 23

H

heavy metal ions, viii, 56, 61, 63, 64, 66, 74, 75, 78, 79, 80, 84, 90, 97, 101, 102, 103, 109, 110
heavy metals, viii, 51, 52, 54, 55, 64, 77, 78, 81, 100, 137
hemicelluloses, viii, ix, 9, 26, 29, 51, 53, 54, 62, 67, 98, 100, 105, 114, 131
hybrid materials, viii, 52, 76
hydrophobic surface layer, viii, 51, 53
hygroscopic, vii, 1, 8

I

ion exchange, viii, 52, 55, 65, 97, 99, 103

J

jute fiber, vii, ix, 1, 2, 3, 4, 5, 6, 7, 8, 9, 10, 11, 13, 14, 16, 17, 18, 20, 21, 22, 23, 24, 25, 26, 27, 28, 29, 30, 31, 32, 33, 34, 35, 36, 51, 52, 53, 54, 55, 56, 57, 58, 59, 60, 61, 62, 63, 64, 65, 66, 67, 68, 69, 70, 71, 72, 73, 74, 75, 76, 77, 78, 79, 80, 81, 82, 83, 84, 85, 86, 87, 89, 90, 91, 92, 94, 95, 98, 100, 101, 102, 103, 104, 105, 106, 107, 108, 109, 110, 111, 113, 114, 115, 116, 117, 118, 119, 120, 121, 122, 123, 124, 125, 126, 127, 129, 130, 131, 132, 133, 134, 135, 136, 137, 139, 140, 141

L

leaf fibers, 114
lignin, viii, ix, 2, 4, 9, 29, 51, 53, 54, 61, 62, 67, 98, 100, 105, 113, 114, 122, 131, 141
lignocellulosic jute fiber, ix, 82, 98, 111
literature review, 54, 98
lubrication, 113, 121

M

manufacturing, vii, 1, 2, 12, 13, 17, 19, 20, 23, 25, 26, 123, 126
manufacturing processes, vii, 1, 2, 12, 13, 17, 19, 20, 23, 25, 26, 123, 126
matrix, vii, 1, 8
membrane filtration, viii, 52, 97, 99
moisture, vii, ix, 1, 8, 14, 17, 26, 28, 29, 30, 32, 37, 42, 87, 93, 95, 113, 114, 118, 122, 134, 135
moisture absorption, vii, 1, 8, 14, 17, 32
multicellular jute fibers, vii, 51

N

natural fibers, vii, 1, 2, 3, 4, 7, 8, 13, 15, 16, 19, 21, 22, 23, 24, 25, 26, 27, 28, 29,

30, 31, 33, 35, 36, 47, 51, 53, 100, 114, 130, 131, 132, 134
naturalistic engineering, 129

O

oxidation, viii, 26, 52, 62, 71, 85, 90, 91, 93, 97, 99, 103

P

pectins, viii, 9, 51, 53, 55, 114
PESTLE analysis, ix, 129, 131, 137, 138, 139, 141
photocatalytic decolorization, viii, 97, 99
piling, 122, 127
pollutants, viii, 51, 52, 53, 54, 67, 75, 76, 77, 86, 93, 97, 99, 100, 104, 106, 107, 109
polyaniline (PANI), 55, 58, 63, 80, 81, 101, 103, 110
polymer, vii, 1, 25, 36
polymer matrix, vii, 1, 2, 8, 10, 20, 35, 134
polymer resins, vii, 1, 25, 36
precipitation, viii, 52, 79, 97, 99

R

raw materials, viii, 19, 52, 76, 100
recyclable, vii, 1, 2, 10, 19, 51, 79, 102, 110, 114
regeneration, viii, 53, 72, 74, 75, 80, 97, 101, 103, 107, 109, 110
research, ix, 11, 12, 16, 17, 21, 22, 30, 33, 36, 37, 38, 39, 40, 41, 42, 43, 48, 64, 70, 71, 74, 82, 83, 84, 88, 90, 111, 127, 129, 132, 140, 141
retting, ix, 104, 113, 114, 115, 116, 117, 118, 126, 127
roving frame, 123

S

saturated adsorbents, 75, 76
seed, 114
seed and fruit fibers, 114

single reed, 117, 118
solid waste, viii, 52, 54, 77
spinning, 3, 44, 113, 115, 118, 119, 122, 123, 124, 126
spinning frame, 123, 124
stability, ix, 3, 11, 18, 19, 23, 26, 43, 90, 98, 107, 113, 121, 138
striping, 117
sustainability, vii, viii, ix, 1, 8, 11, 15, 16, 20, 36, 45, 47, 52, 53, 54, 77, 79, 80, 81, 84, 87, 88, 94, 95, 98, 100, 103, 109, 110, 113, 126, 129, 130, 138, 140, 141
sustainable development, viii, 52, 54, 77, 109, 130
synthetic dyes, 67, 70, 74, 75, 104

T

thermoplastic, vii, 1, 2, 10, 11, 18, 19, 20, 21, 23, 24, 26, 34, 36, 38, 40, 41, 42, 43, 45, 46, 48
thermoplastic polymers, vii, 1, 10, 11
thermosetting, vii, 1, 10, 11, 12, 15, 18, 19, 24, 25
thermosetting resins, vii, 1, 10, 12, 15, 18, 19
treatments, v, vii, viii, 1, 4, 8, 9, 11, 14, 20, 22, 26, 30, 35, 36, 38, 39, 41, 42, 43, 45, 46, 49, 51, 52, 53, 71, 72, 74, 75, 77, 78, 79, 80, 81, 82, 83, 84, 86, 89, 90, 94, 97, 99, 103, 104, 105, 110, 111, 134, 136, 137

V

vegetable fibers, 130, 141

W

wastewater, v, vii, viii, 51, 52, 53, 71, 77, 78, 79, 80, 81, 82, 86, 92, 94, 97, 98, 99, 103, 108, 109, 110, 111
water pollutants, viii, 51, 54, 77, 97, 99
waxes, viii, 51, 53, 131
weaving, 113, 120, 124, 125, 126, 127
winding, 124, 125

Y

yarn, ix, 8, 35, 37, 42, 44, 45, 59, 66, 68, 69, 70, 78, 113, 114, 119, 120, 122, 123, 124, 125, 126, 127

U

UAE, *see* Uterine artery embolization
Ultrasonography, 8
Ultrasound use, 9
Umbilical entry, 129
Unruptured ectopic pregnancy
    diagnosis of, 6
    types of conservative operation, 10
US Food and Drug Administration
    (FDA), 72
Uterine artery embolization (UAE), 35, 37
Uterine tone, 33

V

Ventroscopy, 8

W

Williams Obstetrics, 6, 8

Y

Yolk sac, 24–25